THE DEVELOPMENT OF
SCIENCE-BASED GUIDELINES FOR LABORATORY ANIMAL CARE

Proceedings of the November 2003 International Workshop

Institute for Laboratory Animal Research
Division on Earth and Life Studies

NATIONAL RESEARCH COUNCIL
OF THE NATIONAL ACADEMIES

D0855145

THE NATIONAL ACADEMIES PRESS
Washington, D.C.
www.nap.edu

THE NATIONAL ACADEMIES PRESS • 500 Fifth Street, N.W. • Washington, DC 20001

NOTICE: The project that is the subject of this report was approved by the Governing Board of the National Research Council, whose members are drawn from the councils of the National Academy of Sciences, the National Academy of Engineering, and the Institute of Medicine. The members of the committee responsible for the report were chosen for their special competences and with regard for appropriate balance.

This study was supported by Grant No. RR0118801-01 and Contract No. N01-OD-4-2139, TO 72 between the National Academy of Sciences and the National Institutes of Health. Other contributors were the Association for Assessment and Accreditation of Laboratory Animal Care International, Canadian Council on Animal Care, Centre for Best Practice for Animals in Research (Medical Research Council, UK), Federation of European Laboratory Animal Science Associations, International Council for Laboratory Animal Science, Laboratory Animals, Ltd., and the U.S. Environmental Protection Agency. Any opinions, findings, conclusions, or recommendations expressed in this publication are those of the author(s) and do not necessarily reflect the views of the organizations or agencies that provided support for the project.

International Standard Book Number 0-309-09302-3 (Book)
International Standard Book Number 0-309-54532-3 (PDF)

Additional copies of this report are available from the National Academies Press, 500 Fifth Street, N.W., Lockbox 285, Washington, D.C. 20055; (800) 624-6242 or (202) 334-3313 (in the Washington metropolitan area); Internet, http://www.nap.edu

THE NATIONAL ACADEMIES
Advisers to the Nation on Science, Engineering, and Medicine

The **National Academy of Sciences** is a private, nonprofit, self-perpetuating society of distinguished scholars engaged in scientific and engineering research, dedicated to the furtherance of science and technology and to their use for the general welfare. Upon the authority of the charter granted to it by the Congress in 1863, the Academy has a mandate that requires it to advise the federal government on scientific and technical matters. Dr. Bruce M. Alberts is president of the National Academy of Sciences.

The **National Academy of Engineering** was established in 1964, under the charter of the National Academy of Sciences, as a parallel organization of outstanding engineers. It is autonomous in its administration and in the selection of its members, sharing with the National Academy of Sciences the responsibility for advising the federal government. The National Academy of Engineering also sponsors engineering programs aimed at meeting national needs, encourages education and research, and recognizes the superior achievements of engineers. Dr. Wm. A. Wulf is president of the National Academy of Engineering.

The **Institute of Medicine** was established in 1970 by the National Academy of Sciences to secure the services of eminent members of appropriate professions in the examination of policy matters pertaining to the health of the public. The Institute acts under the responsibility given to the National Academy of Sciences by its congressional charter to be an adviser to the federal government and, upon its own initiative, to identify issues of medical care, research, and education. Dr. Harvey V. Fineberg is president of the Institute of Medicine.

The **National Research Council** was organized by the National Academy of Sciences in 1916 to associate the broad community of science and technology with the Academy's purposes of furthering knowledge and advising the federal government. Functioning in accordance with general policies determined by the Academy, the Council has become the principal operating agency of both the National Academy of Sciences and the National Academy of Engineering in providing services to the government, the public, and the scientific and engineering communities. The Council is administered jointly by both Academies and the Institute of Medicine. Dr. Bruce M. Alberts and Dr. Wm. A. Wulf are chair and vice chair, respectively, of the National Research Council.

www.national-academies.org

INTERNATIONAL WORKSHOP ON THE DEVELOPMENT OF SCIENCE-BASED GUIDELINES FOR LABORATORY ANIMAL CARE PROGRAM COMMITTEE

Hilton J. Klein, VMD (*Chair*), Merck Research Laboratories, West Point, Pennsylvania

Stephen W. Barthold, DVM, PhD, University of California, Davis

Coenraad F.M. Hendriksen, DVM, PhD, Netherlands Vaccine Institute, Bilthoven, Netherlands

William Morton, VMD, National Primate Research Center, University of Washington, Seattle, Washington

Randall J. Nelson, PhD, University of Tennessee Health Science Center, Memphis, Tennessee

Emilie F. Rissman, PhD, University of Virginia Medical School, Charlottesville, Virginia

William S. Stokes, DVM, National Institute of Environmental Health Sciences, National Institutes of Health, Research Triangle Park, North Carolina

Staff

Joanne Zurlo, PhD, Director, Institute for Laboratory Animal Research
Marsha Barrett, Senior Project Assistant
Kathleen Beil, Administrative Assistant
Jennifer Obernier, Program Officer
Susan Vaupel, Editor

Preface

Each country or group of countries addresses the regulation of laboratory animal care in its own way, and even within a single country, there may be different agencies exerting separate regulations or guidelines (e.g., the US has regulations through the Animal Welfare Act administered through the Department of Agriculture and through the Health Extension Act administered through the Department of Health and Human Services as Public Health Service Policy). In Europe, the member nations of the European Union (EU) are governed by Directive 86/609 on the protection of animals used for experimental and other scientific purposes which provides a minimum standard of care for animals in these countries. Each country, in turn, can add more stringent regulations for its own research community, and this has been done. The Council of Europe (CoE), which has 45 member states, adopted the European Convention ETS 123 for the protection of vertebrate animals in 1986. Since the CoE is not a regulatory body, Conventions do not have the force of law, but they do exert a considerable moral pressure, especially in CoE countries for which the Convention is the only pan-European agreement. Nevertheless, once a Member State ratifies a Convention, it becomes a "party" and is bound to be implemented as national law. The standards of housing and care for laboratory animals outlined in Appendix A of ETS 123 served as the basis for these standards in the EU Directive 86/609. These standards are very similar to those specified in the 1996 revision of the *Guide for the Care and*

Use of Laboratory Animals, which was written by an ILAR committee and which serves as the basis of Public Health Service Policy.

The Council of Europe is currently revising Appendix A of ETS 123 during the process known as a Multilateral Consultation, whose participants include representatives from member nations as well as "observers." Nations that are "parties" to the Convention (i.e. have ratified it) play the largest role in acceptance or rejection of the proposed changes recommended by appointed expert working groups for each species or group of species. Observers represent non-member countries, including the US, Canada, and Japan, and non-governmental organizations (NGO), such as the Federation of European Laboratory Animal Science Associations (FELASA), the Eurogroup for Animal Welfare, the European Federation of Pharmaceutical Industries and Associations (EFPIA), and Institute for Laboratory Animal Research (ILAR) of the National Academies (as the authoring body for the *Guide*). The proposed revision of ETS 123 includes changes that may result in substantial differences in recommended housing and care conditions for laboratory animals between Europe and the US. Consequently, many discussions have focused on whether these differences will impact the interchange of research results among countries, and have questioned the value and/or need for harmonization of guidelines among countries.

Arguments may be made for and against harmonization of guidelines for laboratory animal care throughout the world. Intuitively, one might assume that results from studies on animals kept under identical conditions would be more comparable and that harmonization of standards would lead to more collaboration among countries. However, some research has shown that this is not necessarily the case. It is generally agreed that guidelines that incorporate the newest scientific evidence for the best conditions for laboratory animals should also ensure that the data generated are the most reliable. However, if the standards proposed in one country or group of countries are not scientifically based, they are not likely to be freely adopted globally. Financial constraints associated with making major changes could seriously impede the ability to perform animal research at current levels. There is also a concern that in order to avoid making costly changes, institutions may choose to "export" their animal research to countries that have more questionable standards of laboratory animal care, thus generating genuine animal welfare concerns.

There is widespread agreement in the laboratory animal community that these issues need to be continually examined on an international basis. Since ILAR is one of the few organizations with the international reputation and credibility to bring together experts and interested parties from around the world, it was logical that ILAR should host a meeting to discuss these issues. Consequently, an international workshop was held in Washington, DC, in November 2003 to bring together experts from around the world to discuss the available knowledge that can positively influence current and pending guidelines for laboratory animal care, identify gaps in that knowledge in order to encourage future research endeavors, and discuss the scientific evidence that can be used to assess the benefits and costs of various regulatory approaches affecting facilities, research, and animal welfare. This workshop brought together experts from 15 countries over three days to share information, discuss future endeavors, and consider the question of whether or not to harmonize standards. Many fruitful discussions took place during the workshop and the outcome was a better understanding of the cultural influences that serve as a backdrop to regulation and guideline development. The proceedings from this workshop are reported in the pages of this publication.

ILAR wishes to acknowledge and thank the following sponsors of this workshop: the National Institutes of Health (National Center for Research Resources and Office of Laboratory Animal Welfare), Association for Assessment and Accreditation of Laboratory Animal Care International (AAALAC), Canadian Council on Animal Care, Centre for Best Practice for Animals in Research (Medical Research Council, UK), Federation of European Laboratory Animal Science Associations, International Council for Laboratory Animal Science, and Laboratory Animals, Ltd.

Contents

Introduction

Building the Case for Science-based Guidelines—Introductory Remarks

Hilton J. Klein

On behalf of the National Academy of Sciences, the National Research Council, and the Institute for Laboratory Animal Research (ILAR), I would like to extend my appreciation to the speakers and all of the attendees for participating in our meeting over the next several days. I especially thank ILAR Council Chairman Dr. Peter Ward, the members of the ILAR Council, and the members of the ILAR Council's International Committee (Drs. Barthold, Hendriksen, Morton, Nelson, Rissman, and Stokes) for their help in organizing and planning this workshop. It is with sincere gratitude that we thank Dr. Joanne Zurlo, Director of ILAR, Ms. Kathleen Beil, and all of the ILAR staff for their expert guidance and assistance in making this workshop possible today.

It is appropriate to begin this workshop with a quotation attributed to Louis Pasteur: "When our scientific work finds practical application, the cup of joy is overflowing." This quotation sets the tone for our meeting and for all of us who came here to discuss the issues surrounding the need for science-based guidelines in animal-based research. Our goal during this workshop is scientifically to find the optimal way to seek and develop science-based guidelines to serve the needs of the animals and to serve the needs of the scientists. To emphasize one key point, the guidelines must be practical and applied wisely through the use of performance standards, an outcomes-driven approach.

The key question is "What is the issue" that brings us all here? We believe it to be the following. As we move forward and make progress in science and medicine, the issue of regulatory burden on science has emerged as a very important issue for all agricultural research, as well as other types of research. This issue is especially relevant in areas of research where animals are used or somehow involved. The problem of animal research being overburdened with regulations is further accentuated and amplified in complexity by social/cultural issues, animal rights concerns, and pressures from the public sector. This problem faces all of us on an international level. Simply said, the public places high expectations on all of us involved in science to address and solve the problems facing us mutually in medicine, health, and agriculture.

To change my emphasis slightly, we must realize that we face a dilemma. With limitations on government funding (e.g., the National Institutes of Health funding in the United States), the limitations on private industry (e.g., pharmaceutical cost controls), and other economic pressures from competing social, economic, and cultural issues, these issues place a demand on all of us in this room to be more effective and to address the need for more robust output from every scientific laboratory where animals are used in translating their scientific findings into something that has benefit for the public. The reality of the dilemma is very real and very clear. We must be careful as scientists, as laboratory animal specialists, and others to have regulations in place that promote animal welfare and facilitate scientific progress. On the contrary, we must be absolutely certain that we do not promote regulations that do not benefit animals used in research or those that do not facilitate science. In this workshop, I would challenge the laboratory animal experts to think as scientists think and, conversely, to have the scientists place themselves in the position of laboratory animal experts.

There are solutions. We can look to science to help us solve the regulatory burden issue and problem described above. However, the problem can be further presented at several levels.

• First, it is clear that our regulations, standards, and guidelines may not be based on high-quality scientific literature. The quality of studies cited in the literature and the extent of the literature in any particular area may be scant. The extent of this problem varies by species of animal and by area of interest.

• Second, it is clear that there are gaps in our scientific knowledge regarding how best to provide for the welfare of animals in a laboratory setting while facilitating the conduct of good science. We must therefore seek to identify the type of research that will fill the gaps in our knowledge regarding animal welfare issues.

• Third, we must develop regulations, standards, and guidelines that are robust enough to be flexible and dynamic enough to meet the needs of the scientists using the animals and that will stand the test of time.

Ultimately, this effort is critical because this knowledge should allow us to spend our funding and use our limited resources (e.g., facilities and staffing) directly for research, rather than spend it unnecessarily for items the animals may not require. For example, these items may include larger cages, which require more space and decrease capacity, or expensive heating, ventilation, and air-conditioning (HVAC) systems, which are costly to construct and more costly to maintain and/or operate.

THE CHALLENGE

As we seek to define scientifically based regulations and guidelines, we should define the word "science." "Science" is derived from the Latin *scientia* or "knowledge." It is defined in the dictionary as the "systematic observation of natural phenomena for the purpose of discovering laws governing those phenomena." It is somewhat ironic, therefore, that we are using this workshop to define regulations and laws regarding animal use in research based on science. We should examine how regulations, laws, and science can be better linked. This is a good challenge and hopefully one that will be intriguing and enjoyable. The challenge can be broken down further, as follows:

• Identify the gaps in our knowledge in the science behind our regulations and standards.
• Seek to identify the standards that are ideal, or identify the best processes from each country or area of origin.
• Assure ourselves that the rule or regulation making or standard making/development process is not heuristic. That is to say that although representatives of each country or state may believe they have the ideal process or regulation, we must be certain that we do not develop new processes for regulations or guideline development or apply these processes by a simple trial and error method at the international level. More importantly, it must not be a random process.
• Inspection of our processes used for regulation making and guideline development must occur. Inspection, review, and measurement of benefit of these processes will further our ability to identify the best, most effective regulations and standards that benefit animals and are permissive to conducting good science.

- Use science critically and comprehensively to review the literature on which our regulations and guidelines are based. The review can be done by subject area, by species, and so forth.
- Leave behind our other agendas, whether they are political, social, personal, or other, so that there is a dialogue and a meaningful exchange of ideas and information results.
- Focus on ideas to create dynamic, flexible standards that are science based and that benefit animals and science.

The risk of not taking these steps is great and has an impact on all of us. Similarly, the work and effort to change and improve will be great. We are fortunate to have a wealth of expertise present, either with our speakers or with our workshop participants. The knowledge and experience here are significant. I encourage you all to engage and to participate actively. Our speaker sessions, our workshop, and the breakout sessions have been organized to allow you to be participatory and interactive.

Plenary Lecture

Genes, Environments, and Mouse Behavior

John C. Crabbe

GENEOTYPES AND ENVIRONMENTS

As in any area of science, investigators seek to reproduce interesting results of behavioral and other neurobiological experiments with laboratory animals in their own laboratory setting. This generalization of research findings is a crucial part of the scientific process in several ways. Reproducibility, in the broad sense, is taken as a sign of reliability. Failures to reproduce a finding can help to prune the literature of false-positive findings. Successful exportation of a finding to multiple laboratories can allow a scientific insight to be explored using diverse methods not available to the original reporter. In the specific case of studies with stable, reproducible genotypes, the accumulation of results across laboratories is both spatial and temporal. Thus, one of the most long-standing (and reproducible) findings in the modern history of studies with inbred mouse strains is the repeated finding that inbred mice of the C57BL lineage prefer to drink alcohol solutions over plain tap water, and those of the DBA lineage are near-teetotalers, while many other inbred strains show intermediate levels of preference for alcohol (Belknap and others 1993; McClearn and Rodgers 1959; Rodgers 1972; Wahlsten and others 2003a).

However, it is nearly impossible to replicate an experiment exactly. For behavioral studies with laboratory mice, the subject of this paper, it is flatly impossible. Interest in behavioral genetics and genomics is on the rise, driven by the revolution in genomic and informatics capabilities. One of the simplest meaningful behavior genetics experiments with mice is to compare multiple inbred strains on the same task. Within a strain, each same-sex animal is genetically identical to all others, and the individual differences among animals across strains derive from both genetic and environmental sources. When between-strain differences exceed within-strain variability, evidence for significant genetic influence is demonstrated. Because animal husbandry has begun to pay attention to the details of a mouse's genetic background, it is possible to study the same strains on the same behavioral tasks under multiple environmental conditions. Thus, strains might be studied for their activity in a novel arena during their circadian day and night, and/or at different ages, or in a different apparatus. The extent to which mean strain responses on two tasks are correlated may be taken as an estimate of the degree to which a common set of genes influences both traits (Hegmann and Possidente 1981), and such a result would suggest the influence of common neurobiological mechanisms.

The purpose of this presentation is to discuss some examples of the interplay between genotypes and environments, drawn from the behavioral responses of inbred strains of laboratory mice. I start by distinguishing between two broad sources of environmental influences, the *laboratory environment* and the *test environment*. Features of the laboratory environment include (but are certainly not limited to) the local air supply and its humidity, local tap water, noise in the colony rooms, lighting (type, intensity, and light/dark cycle), caging, bedding, food, water delivery system, and all other aspects of husbandry practices. Many of these are unique to a given facility and cannot be exactly duplicated elsewhere (e.g., air), whereas some can be mimicked elsewhere (e.g., food, bedding). Features of the testing environment include the specific apparatus, details of the testing protocols for handling, treating, and scoring the animals, transport to and from colony and home cage, and the specific experimenters performing the work. Testing environments are somewhat more amenable to standardization. The principal point of the paper is to show that strains' behaviors often depend on specifics of the environment. In other words, gene by environment (GXE) interactions occur, even when the exact environmental source of influence cannot be identified.

A MULTISITE TRIAL

Several years ago, my colleagues Doug Wahlsten at the University of Alberta in Edmonton, Bruce Dudek at the State University of New York at Albany, and I set out to evaluate the stability of strain differences in some simple laboratory behaviors. Our principal interest was whether the reliability of the genetic differences on a behavior we saw routinely within each of our laboratories was predictive of reliability of genetic differences across laboratories. After numerous phone calls, meetings, and emails, we decided that one straightforward way to address this question was to standardize the laboratory and test environments nearly completely. We had also often heard during our careers that mice purchased from a supplier "behaved differently" from those reared locally, even when the same inbred genotype was studied. Such complaints were usually accompanied by the certain statement that it was the "stress of shipping" that caused the purchased animals to abandon the true path. We could find no data to support or refute this well-entrenched piece of laboratory lore. We decided to test males and females of eight genotypes in all three laboratories simultaneously on a battery of tests. We further decided to compare directly home-grown mice with those shipped from a breeder.

During the exchange of several hundred emails and more phone calls, we adopted a set of husbandry parameters in common. We purchased the same bedding and food (although the food was from local vendors), and adopted the same light/dark cycle and cage changing schedule. We purchased seven inbred strains and one F1 hybrid as breeding stock at each site, set up matings on the same day, and bred mice locally. We had age-matched mice of each genotype shipped to us for comparison. We built or purchased identical apparatus, adopted exactly the same test protocols, and when the time came, tested 379 mice for activity, elevated plus maze behavior, accelerating rotarod performance, water escape learning, and activity again after a cocaine injection. After the weekend off, mice were given a test of alcohol preference drinking.

The results were largely as we expected, but there were also surprises (Crabbe and others 1999; Wahlsten and others 2003a). For each task excepting time in open arms on a plus maze, by far the most important variable was genotype of the mice. For example, the alcohol preference differences were highly significant, but the only variable that mattered was strain (although, as was also already well known, females drank more than males). The pattern of strain differences was nearly identical in all three laboratories, and it made no difference whether animals were shipped or locally bred. Across all behaviors, the next most important variable was the site at which the test was performed. For example, mice appeared less anxious in the plus maze in Edmonton than at the other sites. Sexes rarely differed, and the effect of shipping was negligible for nearly all variables. However, there were significant GXE interactions for many tests (e.g., the response to cocaine in some strains in Edmonton). Thus, despite a ferocious level of standardization, which amounted to eliminating as much of the environmental variability as possible from the experiment, some strains responded somewhat differently in different laboratories for some tasks.

SOURCES OF INDIVIDUAL DIFFERENCES

The sources of environmental influence that led to strain-specific responses in the multisite trial could not be identified. However, a more recent experiment offers some plausible suggestions. During the early course of his several-year career ranging from postdoctoral fellow to associate professor, Jeff Mogil and his assistants had collected baseline data on a simple, spinally mediated reflex response to acute pain in mice, the tail withdrawal reflex (Chesler and others 2002a,b). Each mouse had its tail immersed in 49°C water, and the latency to remove it was recorded. In fact, 12 different experimenters had amassed data on 8,034 mice from 40 genotypes. Because of the scrupulousness of his laboratory records, he knew age, sex, weight, season of the year, humidity, temperature, cage density, time of day, and order of testing within the cage. He and his collaborator Elissa Chesler hit on the idea of mining this incredible data set to ascertain which variables best predicted individual differences in pain sensitivity. They employed a classification and regression tree (CART) analysis. This automated data-mining technique develops rules used to partition the data recursively. Essentially, it builds "trees," somewhat resembling pedigrees, through successive branch points, serially splitting the data along the most important factors until as much of the variability in the data set as possible has been accounted for. It can be used with unwieldy data sets like this, where there are empty cells and nested factors.

A CART analysis can be used to rank order the factors for their efficacy at explaining individual differences. The most important variable in their outcome was the specific experimenter who performed the experiment. This variable was followed closely by the genotype of the mouse. Other factors that mattered a great deal were season, cage density, and time of day. The other variables were not as important. An attractive feature of this study was that they then obtained 192 new mice from three strains. These mice were tested on the same day, either in the morning or the afternoon, by one of two experimenters. This new experiment revealed the importance of the experimenter, the genotype, and the time of day. In other words, the variables predicted to be important by the CART analysis were verified in an independent study (Chesler and others 2002a,b). It is entirely possible that in the multisite trial, the specific experimenters, who necessarily differed in each laboratory, may have elicited strain-specific responses on certain tasks.

THE BABY AND THE BATH WATER

Does this mean that behavioral genetics is doomed? Are behavioral responses simply too variable, as we often hear from our molecularly inclined colleagues? Is the answer removing the experimenter from the experiment through automation? We tend to disagree with these gloomy thoughts. Rather, we think that the stability of genetic influences is often overlooked. Genotype was the strongest effect for all behaviors in the multisite trial. As Doug Wahlsten and I have continued our work exploring GXE interaction across strains in our two laboratories, we have been studying 21 strains drawn largely from the Mouse Phenome Project A and B list (Paigen and Eppig 2000). We recently explored the literature for evidence for or against stable strain differences in behavior through the years (Wahlsten and others 2003b). We sought tasks where several of the same substrains had been used and where very similar phenotypes were studied, even though apparatus and procedures could not be exactly the same over many years. Thus, we allowed a great deal more environmental variability than we allowed in the multisite trial. For each trait, data had also been collected identically in Portland in 2002. We then correlated the data for older studies with those gathered in Edmonton in 2002.

Another piece of untested laboratory lore is that morphology is less variable than behavior. One trait for which there are many historical data is mouse brain weight. Indeed, in addition to Edmonton and Portland data for 21 strains from 2002, we found eligible studies in 2000, 1973, and 1967. The correlations with Edmonton data for the Portland 2002, 2000, and 1973 studies were all between .84 and .97. These account for 71 to 95% of the variance. However, the oldest study correlated less well with the modern study ($r^2 = 0.23$), though it was based on only four strains. For open field activity, we found studies from another laboratory in 2003, the Portland 2002 data, and studies from 1968 and 1953. All four correlations yielded $r^2 = 0.90$! Clearly, activity in mice is at least as stable across laboratories (and decades) as brain weight, and appears to be more so. The findings were not all so stable however. Although Portland and Edmonton's 2002 elevated plus maze outcomes correlated ($r^2 = 0.78$), a study from 1993 showed only a very modest relationship ($r^2 = 0.37$). because three of the seven strains in common behaved very differently in the two laboratories.

CONCLUSIONS

Understanding complex traits can be advanced through studies with mouse genetic models. However, modeling genetic effects cannot rely on simplistic assumptions about the environment. Although any careful experimenter standardizes conditions within his or her own laboratory to achieve reliable genetic results, it cannot be assumed that within-laboratory reliability translates directly into across-laboratory reliability. Some features of the laboratory environment are nearly impossible to duplicate.

Attempts to standardize the test environment can help improve reproducibility across laboratories, but are not a panacea. Some caveats to enforced standardization of conditions ranging from husbandry to apparatus and protocols should be considered. First, use of a single set of standard conditions could lead to false-negative conclusions. For example, if the effect of a genetically engineered null mutation is not apparent under the standard conditions, and every laboratory adopts them, a real gene effect could be missed. Second, a good deal of time could be wasted exploring apparent gene effects that actually only occur in the standard conditions. Finally, failure to explore a range of environmental conditions may underestimate the actual genetic influence, which is very likely to be expressed as GXE interaction.

ACKNOWLEDGMENTS

These studies were supported by grants from the National Institutes of Health (NIAAA and NIDA) and the Department of Veterans Affairs, and by the National Sciences and Engineering Research Council of Canada.

REFERENCES

Belknap, J.K., Crabbe, J.C., Young, E.R. 1993. Voluntary consumption of ethanol in 15 inbred mouse strains. Psychopharmacology 112:503-510.

Chesler, E.J., Wilson, S.G., Lariviere, W.R., Rodriguez-Zas, S.L., Mogil, J.S. 2002a. Identification and ranking of genetic and laboratory environment factors influencing a behavioral trait, thermal nociception, via computational analysis of a large data archive. Neurosci Biobehav Rev 26:907-923.

Chesler, E.J., Wilson, S.G., Lariviere, W.R., Rodriguez-Zas, S.L., Mogil, J.S. 2002b. Influences of laboratory environment on behavior. Nat Neurosci 5:1101-1102.

Crabbe, J.C., Wahlsten, D., Dudek, B.C. 1999. Genetics of mouse behavior: interactions with laboratory environment. Science 284:1670-1672.

Hegmann, J.P., and Possidente, B. 1981. Estimating genetic correlations from inbred strains. Behav Genet 11:103-114.

McClearn, G.E., and Rodgers, D.A. 1959. Differences in alcohol preference among inbred strains of mice. Q J Stud Alcohol 20:691-695.

Paigen, K., and Eppig, J.T. 2000. A mouse phenome project. Mamm Genome 11:715-717.

Rodgers, D.A. 1972. Factors underlying differences in alcohol preference in inbred strains of mice. In: Kissin B, Begleiter H, eds. The Biology of Alcoholism. New York: Plenum. p. 107-130.

Wahlsten, D., Metten, P., Phillips, T.J., Boehm II, S.L., Burkhart-Kasch, S., Dorow, J., Doerksen, S., Downing, C., Fogarty, J., Rodd-Henricks, K., Hen, R., McKinnon, C.S., Merrill, C.M., Nolte, C., Schalomon, M., Schlumbohm, J.P., Sibert, J.R., Wenger, C.D., Dudek, B.C., Crabbe, J.C. 2003a. Different data from different labs: Lessons from studies of gene-environment interaction. J Neurobiol 54:283-311.

Wahlsten, D., Mosher, T., Crabbe, J.C. 2003b. (In)stability of brain size and behavior over decades in different laboratories. (Abstract). Int Behav Neur Genet Soc Abstr.

Session 1

Current Status:
Identifying the Issues

AAALAC International Perspective

John G. Miller

The Association for Assessment and Accreditation of Laboratory Animal Care (AAALAC) International is a not-for-profit corporation established in 1965 as the American Association for Accreditation of Laboratory Animal Care. From its inception, AAALAC has had sound science as a principal focus. A 1964 report of AAALAC's predecessor, the Animal Care Panel, included the following statements: "As part of the scientific community, the Animal Care Panel has been working to define the conditions of animal care which promote sound and proper animal experimentation. ... The Animal Care Panel cannot and will not proceed with this program [accreditation] without the consent and support of the scientific community." AAALAC's current mission statement continues this emphasis on science, stating that the organization's purpose is to "enhance the quality of research, teaching and testing by promoting humane, responsible animal care and use."

It is most appropriate that ILAR host this meeting, because the animal care and use standard most widely known in the global laboratory animal science community is ILAR's *Guide for the Care and Use of Laboratory Animals (Guide)*. AAALAC is proud to have developed the first edition of the *Guide*, under a 1962 contract from the National Institutes of Health (NIH). Serving as the principal standard used by both AAALAC and the US Public Health Service (PHS) in evaluating animal care and use, each of the seven editions of the *Guide* has been developed by scientists, with its guidance based on "published data, scientific principles, expert opinion and experience with methods and practices that have proved to be consistent with high-quality, humane animal care and use." This hierarchy of scientific support for its recommendations, from peer-reviewed data to experiential evidence, has been a hallmark of the *Guide* and has undoubtedly contributed to its widespread acceptance by the laboratory animal and more general scientific communities. Its utility as an international standard is demonstrated by the fact that the English version has been translated into nine additional languages.

As noted above, the *Guide* is the principal standard used by both AAALAC and the PHS, with both applying its provisions to all vertebrate animals. When one considers the number of animals being used at academic and other institutions that receive support from the NIH and other PHS agencies, and the fact that all major US pharmaceutical companies and commercial suppliers of animals are accredited by AAALAC, it is reasonable to estimate that 90% or more of the research animals in the United States are cared for and used in programs that apply the standards of the *Guide*. This percentage is likely higher for laboratory mice and rats, and refutes the claims of those who state that because the US Department of Agriculture's (USDA's) Animal and Plant Health Inspection Service does not "regulate" mice and rats, these species are not "protected" in the United States.

For nearly the first 15+ years of its existence, AAALAC accredited animal care and use programs only in the United States. Its first accredited program outside the United States was just across the border in Canada, but in 1986 Europe was included with the addition of a program in the United Kingdom. In 1996, the AAALAC Board of Trustees approved a new Strategic Plan that called for significantly increased emphasis on accreditation of programs internationally. Following the directions of that plan has led to remarkable success, with accredited programs currently in 18 countries on five continents. A new AAALAC service, begun in 1997 and called the Program Status Evaluation, has taken AAALAC representatives to additional countries, where institutions are striving to understand the accreditation process and bring their programs up to the AAALAC Standard. Through this international growth, AAALAC has had the opportunity to observe and evaluate programs from the Netherlands to the Philippines, and from Indiana to India. This experience has placed all of us involved with AAALAC, including the Executive Office staff, Council on Accreditation, and ad hoc consultants, in the position of de facto harmonization of different animal care and use standards into the referenced AAALAC Standard above.

AAALAC employs a variety of existing standards and related guidance in its assessment of animal care and use programs. As already mentioned, the *Guide* is our principal standard, and we apply its provisions and principles worldwide. However, the *Guide* is intentionally written in general terms to allow flexibility in its application. Taken together with the *Guide's* emphasis on performance as a measure of successful application, situations naturally occur in which professional judgments regarding appropriate implementation may differ. To assist AAALAC evaluators in these situations and to provide guidance to prospective and current accredited units, we have developed a list of publications and other documents we term our "Reference Resources." The full list is available at http://www.aaalac.org/resources.htm, and includes references from Europe and Canada, in addition to US resources. They provide more specificity than the *Guide* in a wide variety of areas, and in many cases provide examples of appropriate outcomes that are useful when applying the *Guide's* performance standards. The Reference Resources provide guidance in areas such as euthanasia methods *(Report of the AVMA Panel on Euthanasia; Euthanasia of Experimental Animals* (EC DGXI)); training *(FELASA recommendations on the education and training of persons working with laboratory animals: Categories A and C)*; humane endpoints *(Guidance document on the recognition, assessment, and use of clinical signs as humane endpoints for experimental animals used in safety evaluation* (OECD)); and many more.

The *Guide* and AAALAC's Reference Resources share a very important common characteristic—both are science based. The process for adding references to our list requires that the Council on Accreditation vote approval before such addition. The key factor in the Council's consideration of a prospective reference is scientific documentation of its validity and value to an animal care and use program. This factor has led on occasion to the Council's disapproval of proposed resources in which, although originally science based, the supporting data are outdated. Similarly, existing references that have become outdated or have been superceded by newer science-based publications are removed and/or replaced.

Evaluating an entire animal care and use program requires more than the application of the provisions and principles of the *Guide* and Reference Resources. A review of the process by which AAALAC assesses and accredits programs in the United States and internationally also helps demonstrate the mechanisms by which the wide variety of local standards, guidance, and policies are harmonized through the accreditation process to result in a common AAALAC Standard.

In the United States and internationally, the legal and regulatory requirements applicable to the unit being evaluated constitute the baseline for accreditation. No program can receive AAALAC accreditation if it is in violation of the law. Thus, in the United States, all provisions of the Animal Welfare Act Regulations must be met for species covered by the USDA, and for units receiving PHS support, all elements of their Assurance of Compliance with the NIH Office of Laboratory Animal Welfare must also be met. Program elements are then evaluated based on the provisions of the *Guide;* and when necessary and appropriate, specific Reference Resources are used to evaluate performance outcomes in areas in which the *Guide* is nonspecific or institutionally approved deviations from its recommendations have been employed. **Critically important is that all principles of the *Guide* must be met.** Finally, the expert professional judgment of the AAALAC Council on Accreditation is applied through a peer-review process, and a final accreditation status is granted.

Internationally the process is practically identical. Again, no program can become AAALAC accredited if it is in violation of local legal and regulatory requirements. Use of individuals as ad hoc consultants who are familiar with these local requirements facilitates uniform and appropriate interpretation and application to the unit. Once these local baseline requirements are shown to have been met, the *Guide* becomes the next standard to be applied. It is important to note that when local requirements are more stringent than *Guide* recommendations, the former must be met to achieve accreditation. In some instances, the *Guide* includes provisions not addressed in national or supranational animal welfare legislation or regulations, for example, in the area of occupational health and safety. In such cases, two options are available. (1) Other local requirements may exist outside the animal welfare area, as is the case with occupational health and safety requirements in the European Union *(Council Directive on the Introduction of Measures to Encourage Improvement in the Safety and Health of Workers at Work* (Directive 89/391/EEC). (2) In the absence of alternative local standards, the *Guide* standards are used as the basis for evaluating program elements in these areas. As AAALAC has grown internationally, we have conducted assessments in countries without national regulatory standards or other requirements for animal care and use. In these instances—just as in the United States—the *Guide* and Reference Resources serve as the basis for our evaluation. Finally, the application of expert professional judgment through the peer review process by the Council on Accreditation determines a program's final accreditation status. **The key to maintaining consistency and uniformity of the AAALAC Standard across diverse international settings and standards is that all principles of the *Guide* must be met.**

Notwithstanding the broad array of standards and guidance available in the area of animal care and use, circumstances occasionally arise for which there is no applicable published standard. In addition, professional judgments may differ regarding the acceptability of practices or procedures not specifically addressed in existing standards. In these circumstances, AAALAC again looks to science for solutions. In fact, the process used by AAALAC follows that used by ILAR in developing the

Guide, that is, we look for published data in the area in question. When no relevant reports are located, scientific principles and expert opinion form the basis for resolution, with the final decision often informed by Council members' experience with proven methods or practices. An example of this process involves the use of alcohol as a disinfectant. The *Guide* states that "alcohol is neither a sterilant nor a high-level disinfectant," yet it is used extensively for these purposes in rodent survival surgery. To answer questions about the suitability of such use, the Council formed a sub-committee to research and address this issue. Based on the information in six refereed scientific journal articles, two additional references, and the *Manual of Clinical Microbiology,* the Council determined that alcohol was acceptable as a skin disinfectant, but under certain circumstances may not be adequate to sterilize or disinfect surgical instruments. These determinations were published in the AAALAC newsletter, *Connection,* and became part of the AAALAC Standard.

Thus, the AAALAC Standard is not a static document. In fact, it is not based on a single document at all, but rather a compilation of many existing standards, guidelines, and policies that encompass all aspects of an animal care and use program. The majority of these are science based, a fact that not only gives credence to those, like the AAALAC Council on Accreditation, who interpret and apply them in an accreditation program, but also leads to the greater likelihood of acceptance and implementation by the scientists subject to their provisions. The AAALAC Standard is, therefore, an evolutionary product that is developing as internationally recognized standards are interpreted through the collective professional judgment of animal care and use experts and applied through an in-depth, multilayered, scientific peer-review process.

Before providing a list of areas in which I believe more science would be useful, I will comment on the modification of existing standards. It is my strong belief that when existing long-standing guidelines or requirements appear to be meeting the welfare needs of animals, any significant changes should meet three requirements: **(1) the change must be of clear benefit to the animals; (2) it should not interfere unnecessarily with the research; and (3) it should be science based.**

Finally, the following list comprises areas that I believe could benefit from additional scientific study and data. I provide them only as topics for consideration, with the hope that this workshop will serve as a venue for discussion.

1. Enclosure dimensions;
2. Wire-bottom cages;
3. Environmental enrichment;
4. Decapitation/cervical dislocation;

5. Euthanasia in holding rooms;
6. Species separation;
7. Sanitation requirements; and
8. Ventilation requirements.

The Council of Europe: What Is It?

Wim deLeeuw

Shortly after the end of World War II, several movements and activities were born that were dedicated to European unification. As an overall result, the Council of Europe was founded as an international political institution in 1949. It is designed only with international cooperation in mind. The general aims of the Council of Europe are to:

- Protect human rights, democracy, and the rule of law in all member states;
- Promote awareness and encourage Europe's cultural identity and diversity;
- Seek solutions to (social) problems facing European society;
- Consolidate democratic stability in Europe;
- Promote social cohesion and social rights; and
- Promote and develop a European cultural identity with emphasis on education.

The actual areas of concern are human rights, health, education, culture, youth, sport, the environment, local democracy, heritage, legal cooperation, bioethics, animal welfare, and regional planning. Today, the Council of Europe has 45 member states, including about 800 million people.

The Council of Europe must be distinguished from the European Union, which was set up in 1957 as the European Economic Community. First, it is not a supranational institution like the European Community. It does not have legislative power. Its member states are cooperating on a voluntary basis. The Council of Europe cannot impose any rule on its member states. Second, unlike the European Union, the Council of Europe is not an economic organization.

The geographical confines of the Council of Europe are larger than the membership of the EU. The EU has 15 member states: Austria, Belgium, Denmark, Finland, France, Germany, Greece, Ireland, Italy, Luxembourg, the Netherlands, Portugal, Spain, Sweden, and the United Kingdom. All of these countries are also member states of the Council of Europe. Next to these member states, however, the Council of Europe also includes 30 other European countries: Albania, Andorra, Armenia, Azerbaijan, Bosnia and Herzegovina, Bulgaria, Croatia, Cyprus, the Czech Republic, Estonia, Georgia, Hungary, Iceland, Latvia, Liechtenstein, Lithuania, Malta, Moldova, Norway, Poland, Romania, the Russian Federation, San Marino, Serbia and Montenegro, Slovenia, Slovakia, Switzerland, the former Yugoslav Republic of Macedonia, Turkey, and Ukraine.

THE COUNCIL OF EUROPE: HOW DOES IT WORK?

The headquarters of the Council of Europe, Le Palais de l'Europe, is situated in Strasbourg, France. The Committee of Ministers is the decision-making body of the Council of Europe. It is composed of the Ministers of Foreign Affairs of the member states. This body officially adopts Conventions, Resolutions, Agreements, and Recommendations. The Committee of Ministers also ensures that the conventions and agreements are implemented. In addition, there are two other institutions: (1) The Parliamentary Assembly is the organization's deliberative body, the members of which are appointed by national parliaments. (2) The Congress of Local and Regional Authorities of Europe is a consultative body that represents local and regional authorities. Governments, national parliaments, and local and regional authorities are thus represented separately at the Council of Europe level.

The main tools of the Council of Europe to achieve its objectives are the following legal instruments:

• Recommendations—often referred to as "soft law." There is no legal obligation to follow or implement these recommendations; and

• Conventions or treaties concluded between states. The member states are not legally obliged to sign a Convention, although they may be expected to do so since under the Council of Europe's Statute, they have undertaken to "collaborate sincerely and effectively in the realization of the aim of the Council." Nonetheless, there are different ways a member can deal with a Convention. It may choose to ignore the Convention as being not relevant or not applicable to the national situation. By taking that position, a member is not obliged to comply with its provisions. A member can sign the Convention, thus recognizing the value and existence of the Convention. After having signed a Convention, a member is still not obliged to comply with the provisions of the Convention. However, once a state has signed and ratified (i.e., its Parliament has approved the instrument) and the Convention has become effective, the state will be morally and legally bound under international law to implement the Convention. Thus, the state has become a Party to that Convention and must ensure that the provisions will be respected on its territory. Most Council of Europe Conventions are not directly applicable within a member state; they are not "self-executing." The most common way for a state to implement them is to enact appropriate national legislation or to adapt its existing domestic law to make it correspond to the rules in the Convention.

In contrast to the European Union, practically spoken, there is little legal enforcement of Conventions, which probably leads to variability in compliance. Some Conventions are also open for adoption by nonmember states. The Conventions and recommendations are drafted by governmental experts responsible to the Committee of Ministers, thereby providing for the interaction of political interests with technical considerations. They only have a legal status after they are adopted by the Committee of Ministers.

More than 350 nongovernmental organizations (NGOs) are granted a consultative status. Within the context of the Council of Europe, there are several consultation arrangements, which enable NGOs to participate in intergovernmental activities and encourage dialogue. These NGOs are a vital link to the public at large and to specific parts of society.

THE COUNCIL OF EUROPE AND ANIMAL WELFARE

The work of the Council of Europe on animal protection was started in the 1960s. Since then, the following five Conventions on the protection of animals have been drawn up:

(1) On animals during international transport (ETS 65, 1968), which establishes general conditions for the international transport of animals;

(2) On animals kept for farming purposes (ETS 87, 1976), which is a framework convention. More detailed recommendations on species are given in separate guidelines;

(3) On animals for slaughter (ETS 102, 1979);

(4) On vertebrate animals used for experimental and other scientific purposes (ETS 123, 1986); and

(5) On pet animals (ETS 125, 1987).

All of these Conventions are based on the principle that "for his own well-being, man may, and sometimes must, make use of animals, but that he has a moral obligation to ensure, within reasonable limits, that the animal's health and welfare [are] in each case not unnecessarily put at risk."

These Conventions were the first international legal instruments to establish ethical principles for the use and handling of animals. They are the result of very lengthy research, discussions, and negotiations, undertaken by governmental experts, delegates from animal welfare organizations, scientific researchers, and representatives of professional associations directly concerned. They are therefore the results of compromises. The political and technical value of the legal instruments working method adopted in the framework of these activities is based on the close collaboration between representatives of all the governmental and nongovernmental organizations that are involved. They have been used as a basis for, and continue to influence, all of the national relevant legislation in Europe.

THE COUNCIL OF EUROPE AND THE PROTECTION OF LABORATORY ANIMALS

As early as 1971, the parliamentary assembly recognized that to protect animals against abusive and unnecessary experimentations certain norms should be established at an international level, to enable states to regulate such experiments in an harmonious way in their domestic law. A first draft of the Convention was elaborated by the Ad Hoc Committee of Experts on the Protection of Animals, the CAHPA. After lengthy discussions, the Convention was finally adopted in May 1985. The Convention is accompanied by an explanatory report, and attached to it are technical appendices. Appendix A presents guidelines for the accommodation and care of animals. Existing German and US guidelines were used as a basis. Unlike the provisions of the Convention itself, the guidelines in Appendix A are not mandatory; they are recommendations. These guidelines are based on knowledge of that time and good practice. Appendix A explains and supplements the principles on accommodation and care as adopted in article 5 of the Convention. Appendix B contains tables for the presentation of the statistical data on the use of animals for experimental and other scientific purposes. The object of the Appendix is thus to help authorities, institutions, and individuals in their pursuit of the aims of the Council of Europe in this matter.

The European Convention for the protection of vertebrate animals used for experimental and other scientific purposes (1986, ETS 123) includes provisions concerning the scope, care, and accommodation of the animals, conduct of experiments, humane killing, authorization procedures, acquisition of animals, control of breeding or supplying and user establishments, education and training, and statistical information. It is clearly visible from several provisions that the 3Rs of Russell and Burch are used as a basis for the Convention.

Currently, 15 countries have signed and ratified ETS 123 and thus are Parties to the Convention: Belgium, Cyprus, Czech Republic, Denmark, Finland, France, Germany, Greece, the Netherlands, Norway, Spain, Sweden, Switzerland, the United Kingdom, and the European Community. The Convention is signed by Bulgaria, Ireland, Portugal, Slovenia, and Turkey. The Convention provides for Multilateral Consultations of the Parties at least every 5 years, to examine the application of the Convention and the advisability of revising it or extending any of its provisions according to changes of circumstances and new scientific evidence. The Multilateral Consultations are prepared by a Working Party. For their work, the Parties have invited other member states of the Council of Europe and nonmember states and cooperate very closely with nongovernmental organizations that represent the fields concerned. In the preparatory meetings for the 4th Multilateral Consultation, the following observers participated:

Canadian Council on Animal Care (CCAC)
European Biomedical Research Association (EBRA)
European Federation of Animal Technologists (EFAT)
European Federation for Primatology (EFP)
European Federation of Pharmaceutical Industries and Associations (EFPIA)
European Science Foundation (ESF)
Federation of European Laboratory Animal Breeders Associations (FELABA)
Federation of European Laboratory Animal Science Associations (FELASA)
Federation of Veterinarians of Europe (FVE)
International Council for Laboratory Animal Science (ICLAS)
Institute for Laboratory Animal Research (ILAR)
International Society for Applied Ethology (ISAE)
World Society for the Protection of Animals (WSPA)
Eurogroup for Animal Welfare (Eurogroup)
Member States: Austria, Croatia, Hungary
Nonmember State: United States of America

The participation of representatives of observer states and nongovernmental organizations is of great value. It implies a very broad exchange of information at technical as well as legal and political levels. Therefore, their involvement in this work has to be associated with the success of the Multilateral Consultations to ensure a common and satisfactory level of protection for animals used for scientific purposes, thus enabling the Council of Europe to maintain its position of initiator in Europe for the protection of these animals.

Until now, three Multilateral Consultations have been held. At the 1st Multilateral Consultation held in 1992, the Parties adopted a resolution in which the scope of the Convention was made more precise in respect for genetically modified animals, and certain tables for statistical data were remodeled.

At the 2nd Multilateral Consultation that was held in 1993, a resolution on education and training of persons working with laboratory animals was adopted. This resolution contained guidelines for the education and training of persons taking care of animals (Cat. A), persons carrying out procedures (Cat. B), and persons responsible for directing or designing procedures and animal science specialists (Cat. D). The guidelines included in the resolution were mainly based on a report that had been issued by FELASA.

At the 3rd Multilateral Consultation that was held in 1997, a resolution on the acquisition and transport of animals was adopted. This resolution contained guidelines that complemented the guidelines on this topic included in Appendix A.

Concerning the care and accommodation of animals, the Parties recognized that Appendix A had proven to be of great value and was widely used as a reference. At the same time, however, it was realized that the Appendix had been drafted more than 10 years ago. The Parties agreed that new scientific evidence and new experience since then made it necessary to revise the Appendix and to define the areas where further research is needed. They therefore agreed that this revision of Appendix A should be on the agenda of the 4th Multilateral Consultation. Pending this revision, a resolution was drafted presenting guidelines for the improvement of the accommodation and care of laboratory animals, which would complement the guidelines in Appendix A. The guidelines in the resolution were mainly based on the conclusions and recommendations of the International Expert Workshop on laboratory animal welfare that was held in 1993 in Berlin. It was concluded that the most important areas appeared to be the enrichment of the environment of the individual species according to their needs for the following:

- Social interaction. Group or pair housing was considered to be preferable to individual housing for all gregarious species, as long as the groups are stable and harmonious;
- Activity-related use. Cages should be structured to enable an activity-related use of the space available; and

• Appropriate stimuli and materials. It was recognized that guidelines could never replace close and regular observations of the animals involved to ensure that the enrichment initiatives do not have adverse effects for groups or individuals.

Taking into account the evolution of scientific knowledge and changing circumstances, the Parties realized that the technical Appendices might need to be adapted more frequently than its main provisions. However, because these Appendices are an integral part of the Convention, such adaptations could result in complicated amendment procedures. Therefore, a Protocol of Amendment (ETS 170) providing for a simplified procedure for the amendment of the technical Appendices to the Convention was drafted and opened for signature in June 1998. Thus, the Parties are able to amend the technical Appendices, without formal adoption by the Committee of Ministers.

The finalized documents must be formally adopted at the 4th Multilateral Consultation. Thereafter they will be submitted to the Committee of Ministers. The text of the Convention and the related documents, such as resolutions adopted by the Committee of Ministers, as well as the draft proposals for the revision of the Appendix on which the discussion is finalized and the finalized background documents, are available on the website of the Council of Europe (www.coe.int/animalwelfare/).

The work that has been done at the Council of Europe in the area of laboratory animal welfare was based on a very fruitful cooperation between member states and observers of various organizations. To be more effective, it will be very important that the cooperation between the European Union and the Council of Europe is intensified and that cooperation with other international umbrella organizations is developed further.

ICLAS and the International Community

Gilles Demers

HISTORY

Through an initiative of the United Nations Educational, Scientific, and Cultural Organizations (UNESCO), the Council for International Organizations of Medical Sciences (CIOMS), and the International Union of Biological Sciences (IUBS), the International Committee on Laboratory Animals (ICLA) was conceived in 1956 as a nongovernmental organization to promote high standards of laboratory animal quality, care, and health. Its activities have included collaboration with the World Health Organization since 1961. In 1979, ICLA was renamed the International Council for Laboratory Animal Science (ICLAS), because much new knowledge in biology and medicine requires planned experiments with organisms or their parts.

ICLAS is an international nongovernmental and nonprofit scientific organization. ICLAS exists to promote high standards of animal care and use in education, research, testing, and diagnosis, to promote good science and foster humane practices in scientific research. The ICLAS Mission and Aims are compatible with the highest possible standards of animal research internationally.

MISSION AND AIMS

ICLAS advances human and animal health by promoting the ethical care and use of animals in research worldwide. The aims of ICLAS are

- To promote and coordinate the development of laboratory animal science throughout the world and as a matter of priority in developing countries;
- To promote international collaboration in laboratory animal science;
- To promote quality definition and monitoring of laboratory animals;
- To collect and disseminate information on laboratory animal science;
- To promote worldwide harmonization in the care and use of laboratory animals;
- To promote the humane use of animals in research through recognition of ethical principles and scientific responsibilities; and
- To promote the 3Rs tenets of Russell and Burch.

MEMBERSHIP

ICLAS is composed of four (4) categories of members: National members (30); Scientific/Union members (37); Associate members (34); and Honorary members (9). National members represent national perspectives. Scientific/Union members represent national or regional laboratory animal science and other scientific associations. Associate members represent commercial and academic organizations that support the aims of ICLAS.

List of ICLAS Members

National members: Argentina, Austria, Belgium, Canada, China, Costa Rica, Cuba, Cyprus, Denmark, Finland, Germany, Hong Kong, Hungary, India, Iran, Ireland, Israel, Italy, Japan, Netherlands, Norway, Poland, South Africa, Spain, Sweden, Thailand, the United Kingdom, the United States, Mexico, and Tunisia.

Scientific/Union members: AALAS/USA, ANZLAS/Australia-New Zealand, AGS/USA, CALAS-ACSAL/Canada, CALAS/China, GV-SOLAS/Germany, JALAS/Japan, KALAS/Korea, LASA/U.K., NVP/Netherlands, Scand-LAS/Sweden, SEEA/Spain, AFSTAL/France, CSLAS/Taiwan, SECAL/Spain, FinLAS/Finland, BCLAS/Belgium, Balt-LAS/Latvia, ACCMAL/Central America, AAALAC, ACLAM/USA, CSLAS/Croatia, KRIBB/Korea, JSP/Japan, SGV/Switzerland, AACyTAL/Argentina, AMCAL/Mexico, COBEA/Brazil, BALAS/Bangladesh, TALAS/Thailand, International Union of Biological Sciences (IUBS)/France,

International Union of Immunological Societies (IUIS)/Netherlands, International Union of Nutritional Sciences (IUNS)/Netherlands, International Union of Pharmacology (IUPHAR) Germany, International Union of Physiological Sciences (IUPS)/USA, and International Union of Toxicology (IUTOX)/Switzerland.

NEW ICLAS GOVERNING BOARD (2003-2007)

President: Gilles Demers (Canada)
Vice President: Norikazu Tamaoki (Japan)
Secretary-General: Patri Vergara (SECAL, Spain)
Treasurer: Cecilia Carbone (Argentina)

National Members

Gemma Perretta (Italy)
Czeslaw Radzikowski (Poland)
Norikazu Tamaoki (Japan)
Guy De Vroey (Belgium)

Scientific/Union Members

Denna Benn (CALAS, Canada)
Melvin Dennis (AALAS, USA)
Guy Dubreuil (AFSTAL, France)
J.R. Haywood (IUPHAR, USA)
Rafael Hernandez (ACCMAL, Mexico)
Toshio Itoh (JALAS, Japan)

STRATEGIC PLAN

The ICLAS Governing Board has developed a Strategic Plan to guide the organization through the next several years. The Strategic Plan includes the mission statement of ICLAS: "The International Council for Laboratory Animal Science advances human and animal health by promoting the ethical care and use of animals in research worldwide." ICLAS strives to act as a worldwide resource for laboratory animal science knowledge; to be the acknowledged advocate for the advancement of laboratory animal science in developing countries and regions; and to serve as a premier source of laboratory animal science guidelines and standards, and as a general laboratory animal welfare information center.

ICLAS PROGRAMS

Meetings

International meeting. An international scientific meeting is held in association with the general assembly every 4 years. It is organized by a National or Scientific member and is often held in association with regional/local organizations.

Regional meetings. Other regional scientific meetings and courses are organized by laboratory animal science organizations in the various regions of the world under the auspices of six ICLAS Regional Committees for the following regions: Europe, Asia, Africa (French and English regions), Oceania, and the Americas. This process has allowed ICLAS to focus on each region and to assure diffusion of scientific knowledge within all regions of the world. ICLAS provides funding and guidance for courses and meetings in these regions.

Communications

ICLAS FYI Bulletin. The ICLAS FYI Bulletin is an electronic instrument that provides worldwide distribution of timely information that may be of interest to ICLAS constituents and that may be passed on to their constituents. An average of five bulletins are sent each month, and these bulletins have led to interaction among laboratory animal scientists around the world. This international network is the most extensive in laboratory animal science in the world.

ICLAS Website: www.iclas.org. The ICLAS web page has been developed to provide various items of information on ICLAS programs and ICLAS activities. This information is important to existing and potential constituents.

ICLAS INITIATIVES

ICLAS-CCAC International Symposium on Regulatory Testing and Animal Welfare (Québec, Canada, June 2001)

This ICLAS initiative was a great success, and included 160 participants from 22 countries. The proceedings of this meeting were published in *ILAR Journal* in the fall of 2002, and included the following conclusions:

- A definite link between good animal welfare and quality science;
- Reduction of pain and distress is a higher priority than the reduction of numbers of animals;

- Guidance documents on humane endpoints: CCAC (1998) and OECD (2000) guidelines recognized as effective refinement tools to minimize pain and distress;
  - Existing validated earlier endpoints should be used by all sectors;
- Guidelines developed by the OECD and ICH to promote more humane methodologies for the testing of chemicals are reducing animal use by eliminating redundant testing;
  - Data sharing and training programs should be put in place quickly to assist regulators, toxicologists, and others to be comfortable with the new tests;
- Animal care practices that improve animal welfare without jeopardizing the scientific design must be implemented.

ILAR International Workshop on Development of Science-based Guidelines for Laboratory Animal Care

ICLAS is a cosponsor of this workshop convened to discuss the available knowledge that can affect current and pending guidelines for laboratory animal care, identify gaps in that knowledge to encourage future research endeavors, and discuss the scientific evidence that can be used to assess the benefits and costs of various regulatory approaches affecting facilities, research, and animal welfare.

Meeting for Harmonization of Guidelines (FELASA 2004, France)

The harmonization of existing guidelines for the use of animals in research, teaching, and testing is an emerging issue in the context of the globalization of research around the world. ICLAS, as an international umbrella organization, could act as a facilitator in this area. Accordingly, ICLAS will be inviting one or two representatives of the principal organizations in the world that produce or use guidelines for the use of animals in research, to a 1-day meeting (June 13-14, 2004) held in conjunction with FELASA 2004, in Nantes, France. Representatives from ILAR, FELASA, CCAC, Council of Europe, OECD, ICH, AAALAC, and others will be invited. This will be an opportunity to open the dialogue on harmonization of some existing guidelines and to learn whether there are possibilities to reach a consensus on the recognition of these guidelines at an international level. According to the commitment of the participants, this initiative could be repeated on a regular basis.

CONCLUSION

In the context of the ILAR International Workshop on Development of Science-based Guidelines for Laboratory Animal Care, ICLAS can play an important role because of

1. Its role as an international umbrella organization:

a. ICLAS membership includes countries from every region of the world; and

b. The impact of ICLAS programs to ensure diffusion of good science and good animal welfare practices is felt worldwide, through ICLAS Meetings, Regional Programs, the Communication Program, and other ICLAS Initiatives.

2. The ICLAS Policy regarding harmonization versus standardization.

ICLAS supports the harmonization of animal care and use policies, guidelines, and other forms of regulation on a worldwide basis. However, because ICLAS is in constant liaison with countries and regions having different cultures, traditions, religions, legislations, regulations, and laws, ICLAS considers that each country should be able to maintain an animal welfare oversight system that reflects those elements and that suits the country's own particular characteristics. The rigidity related to standardization for all does not fit with respect to the characteristics of individual countries.

Role of the National Institutes of Health Office of Laboratory Animal Welfare and the Public Health Service Policy on Humane Care and Use of Laboratory Animals

Nelson L. Garnett

The laws, regulations, and policies in the United States have three main sources of which two are government and one is a private voluntary accreditation body: (1) the US Department of Agriculture (USDA), (2) the Department of Health and Human Services, and (3) the American Association for Assessment and Accreditation of Laboratory Animal Care International (AAALAC). In this presentation, I will focus mostly on the Public Health Service (PHS) contribution, although there are many similarities among the three sources.

I am responsible for a portion of the US system. To reiterate, we start with a law, the Public Health Service Act. It is the same law that authorized the establishment of the National Institutes of Health (NIH) many years ago. The PHS Act was amended in 1985 to include the PHS Policy on Humane Care and Use of Laboratory Animals, implemented by my office, the Office of Laboratory Animal Welfare (OLAW). It is interesting to note that the PHS Policy preceded the legislation that authorized it, so it was ready to be implemented almost immediately after passage. We believe that Congress was very familiar with the contents of the PHS Policy and provided a strong endorsement by authorizing it virtually unchanged.

The latest version of the PHS Policy was reprinted in 2002 with minor changes to reflect updated references, addresses, and a name change for our office. Otherwise, it is unchanged from the 1986 version. I will provide the websites where you can download this and many other relevant documents in the discussion. The PHS Policy spells out the requirements for animal care that must be followed by all institutions wishing to be eligible to receive funding from any of the PHS agencies. The PHS Policy also requires that institutions receiving PHS support design their programs to conform with the *Guide for the Care and Use of Laboratory Animals* (the *Guide*).

You have already heard much about the *Guide* as it relates to AAALAC, but it has always been linked to compliance with the PHS Policy as well. One more example of the harmonization of the regulations and policies is the fact that almost all of the institutional animal care and use committee (IACUC) procedures in the USDA regulations were drawn from the PHS Policy. Many of the existing USDA cage size requirements were taken verbatim from the earlier version of the *Guide*.

OLAW is part of the NIH, the primary government funding source for biomedical research. My immediate boss is responsible for the arm of NIH that funds billions of dollars worth of research at universities and other institutions in the United States and abroad. About half of that research includes some animal component. OLAW is composed of three major components: Assurances, Compliance, and Education. The total staff has recently grown to 10, including seven professionals and three support staff. We are responsible for monitoring approximately 1000 institutions that receive PHS support.

The applicability of the PHS Policy includes all animal-related activities conducted or supported by the PHS. The main PHS agencies involved in funding animal research are the NIH, the Centers for Disease Control and Prevention (CDC), and the Food and Drug Administration (FDA). This coverage is very broad and includes intramural and extramural research, grants and contracts, subcontracts, training grants, cooperative agreements, domestic and foreign activities, and even some collaborations and purchase orders. It applies to all live vertebrate animals supported by the PHS, without exception. In addition to the PHS agencies, many others such as the National Science Foundation, Howard Hughes Medical Institute, and American Heart Association have adopted the PHS Policy for their own funding programs. These endorsements have greatly expanded the influence of the policy beyond its original intent.

All PHS-supported activities must be conducted at an assured institution and must be reviewed and approved by an IACUC. PHS awarding units are responsible for ensuring that these requirements have been met before considering proposals for funding. Each grant applicant must address "Five Points" within the body of the application before the scientific peer review. Applications without this information are considered "incomplete." These points are number and space rationale descriptions of use procedures to minimize pain methods of euthanasia.

The US Government Principles (at the back of both the PHS Policy and the *Guide*) provide the foundation for all federal regulations regarding animals, and the PHS Policy was expressly written to implement these principles. I should also mention that these principles are virtually identical to the International Guiding Principles, which are applicable worldwide. It is also important to note that these principles apply not only to biomedical research, but also to *testing* as well as *training* (teaching).

Briefly, OLAW is responsible for the implementation and interpretation of the PHS Policy, the negotiation and approval of Assurances, the evaluation of noncompliance, and a nationwide education program. The current PHS Policy was already in existence and ready to be put into place almost immediately after the legislation that passed in 1985 had mandated it. It was patterned after the widely accepted human subjects protections that were implemented out of the same office until several years ago. The key elements of the PHS Policy include the philosophy of enforced self-regulation, an Assurance mechanism with oversight by a local institutional committee, and appropriate reporting and documentation. The policy applies to all PHS-supported activities and covers all vertebrate animals.

The PHS concept of enforced self-regulation includes a reliance on performance standards wherever possible, and recognizes the need for flexibility and professional judgment. To be effective, it must be self-monitoring, self-correcting, and self-reporting.

WHAT IS AN ANIMAL WELFARE ASSURANCE?

The Assurance is a written document that provides the basis for a trust relationship between the institution and the government. It describes your unique program of animal care and use and, once approved, becomes a criterion for future evaluation. An approved Assurance is required for eligibility to receive PHS support. The Assurance describes, in some detail, the program of animal care and use and must address all of the following elements of that program: applicability, lines of authority and responsibility, the IACUC, procedures to implement PHS Policy, veterinary care, occupational health, personnel qualifications, facilities, and species. The Assurance must be signed by the Institutional Official, someone in the institution who is authorized to make commitments on behalf of the institution and to ensure that the conditions of the PHS Policy are met.

Many of you are familiar with the role of the IACUC in protocol review and facility inspections. What is not always understood is that the IACUC is usually advisory to, or acting on behalf of, the Institutional

Official in carrying out its duties. Two other responsibilities that do not always receive the attention they deserve are the review of *programs* and the investigation of animal welfare concerns.

Protocol review procedures include the following:

- Review of all animal-related activities;
- Provision for designated reviewer(s) to conduct a review (after all members have had an opportunity to call for a full review);
- Ability to consult experts if needed;
- Requirement to review all significant changes before their initiation; and
- Monitoring ongoing activities (perhaps one of the most frequently overlooked responsibilities).

The IACUCs are expected to take the following considerations:

- Animal procedures involving analgesia and anesthesia, euthanasia;
- Environmental conditions;
- Veterinary medical care;
- Personnel qualifications.
- Specific USDA requirements to verify that (1) proposals avoid unnecessary duplication, and (2) the Principal Investigator has considered alternatives to painful procedures.
- Review of certain broad general elements of "the science" (e.g., relevance to human/animal health, advancement of knowledge, good of society, species and numbers, and consideration of nonanimal methods). IACUCs are not expected to conduct scientific peer review, but instead to consider certain basic ethical issues as inseparable from the science.

Every 6 months, the IACUC is required to conduct animal facility and program evaluations. Reports of these evaluations are submitted to the Institutional Official and describe the institution's adherence to, or departures from, the *Guide*. Reasonable and specific plans and schedules are then developed for correcting those deficiencies.

Annual reporting is considered vital to OLAW's ability to provide oversight. These reports are to include the dates of semiannual evaluations, significant program or facility changes, change in accreditation status, and changes in IACUC membership; and they must allow for minority views to be expressed. Prompt self-reporting of problems is an essential part of our trust relationship. The majority of compliance issues with which we deal are brought to our attention by the institutions themselves. We view the reporting of problems, along with the corrective actions taken, to be a positive sign that the system is working as expected.

Finally, I would like to add a few words about the most important ingredient in our animal welfare oversight system—education. Education is the preventive medicine in our business. OLAW cosponsors an entire series of animal welfare educational activities throughout the year, with active participation from all of the other players, including the regulated community. The wealth of information is beyond the scope of this presentation, but I invite you to visit this website (http://grants.nih.gov/grants/olaw/olaw.htm) and explore the many resources and links that it has to offer.

Regulatory Authority of the US Department of Agriculture Animal and Plant Health Inspection Service

Chester A. Gipson

Congress passes, and the President signs, all legislation authorizing activities of the Animal and Plant Health Inspection Service (APHIS) of the US Department of Agriculture (USDA). The laws authorize or direct the Secretary of Agriculture to take certain actions, which may include issuing regulations. The Secretary delegates authority to the Under Secretary of Marketing and Regulatory Programs (MRP), who then delegates authority to the Administrator of APHIS. The Administrator of APHIS delegates authority to the APHIS Associate Administrator and Deputy Administrators/Directors. The Animal Care (AC) program of APHIS receives its regulatory authority from the Animal Welfare Act (7 U.S.C. 2131-2159) and the Horse Protection Act (15 U.S.C.1821-1831).

ADMINISTRATIVE PROCEDURE ACT

The Administrative Procedure Act (APA) contains the basic requirements for federal rulemaking. For most rulemaking, the APA requires the following:

- Publication in the *Federal Register* of a proposed rule, including either the terms or substance of the proposed rule;
- Opportunity for public participation in rulemaking through submission of written comments on the proposed rule;
- Publication in the *Federal Register* of a final rule, with an explana-

tion of any changes that the agency has made and a response to the public comments; and

• An effective date for the final rule that is at least 30 days after publication in the *Federal Register*, unless the rule relieves restrictions, grants an exemption, or there is other good cause for making an exception. This kind of rulemaking is called "informal" or "notice and comment" rulemaking.

Publication of a rule in the *Federal Register* has certain legal effects. The rule provides official notice of the existence and content of a document. Publication indicates that the document was issued properly. Finally, publishing the rule provides evidence that it is judicially noticed by a court of law. Regulations that are not published in the *Federal Register* in accordance with the Administrative Procedure Act may not be upheld in a court of law. Therefore, any rules that an agency wishes to enforce should be published in the *Federal Register*.

OTHER ACTS AND EXECUTIVE ORDERS

Although the APA sets forth the basic requirements for federal rulemaking, other acts and executive orders also apply and include the following: (1) Executive Order 12866, which provides for review of federal rules by the Office of Management and Budget (OMB) (part of the Office of the President) and which requires the preparation of cost-benefit analyses for some rules; and (2) the Regulatory Flexibility Act, which requires analyses by agencies of the potential economic effects of their rules on small entities (small businesses, nonprofits, and small governmental jurisdictions).

TYPES OF RULES

Among the several types of rules are a proposed rule, a final rule, an interim rule, an advance notice of proposed rulemaking, and a direct final rule. Each type is described briefly below.

Proposed Rule

Most rulemaking in APHIS begins with a proposed rule. This document must contain a preamble that includes the following, at a minimum: an explanation of the proposed rule; an analysis of the anticipated economic effects of the proposed rule; a description of any information collection requirements; an invitation to the public to submit comments by a specified date (usually 60 days after publication); and the proposed rule itself, as it would appear in the Code of Federal Regulations.

Final Rule

Most rulemaking in APHIS concludes with a final rule. This document must contain a preamble that includes the following, at a minimum: a response to the issues raised by the public comments; an analysis of the anticipated economic effects of the final rule; a statement concerning any information collection requirements contained in the rule; and an effective date for the rule. The effective date must be at least 30 days after publication, unless the final rule relieves restrictions or there is other good cause for making the rule effective sooner. The final rule document must also contain the rule text that will appear in the Code of Federal Regulations.

Interim Rule

An interim rule may be issued instead of a proposed rule when there is good cause for making a rule effective before the public has an opportunity to comment upon it. For example, APHIS may need to put immediate restrictions in place after an outbreak of an animal disease to prevent the spread of that disease. An interim rule may be followed by a final rule, which could contain changes to the interim rule based on public comments. If the final rule does not make any changes to the interim rule, APHIS calls the final rule an affirmation of the interim rule. An interim rule contains a preamble that must include the following, at a minimum: an explanation of the rule; an effective date (usually upon publication); a description of any information collection requirements and the emergency approval number from the OMB necessary for implementing them; and an invitation to the public to submit comments by a specified date.

Advance Notice of Proposed Rule

Yet another type of rulemaking document is the advance notice of proposed rulemaking. This type of document may be used when the agency seeks to obtain preliminary information before issuing a proposed rule, or even making a decision about whether to issue a proposed rule. This document contains a description of the rulemaking being considered; an invitation to the public to submit comments by a specified date; and specific questions or issues that APHIS believes the public should address.

Direct Final Rule

The direct final rule is a type of rule that provides a shortcut for noncontroversial rules that are unlikely to generate even one negative comment. The direct final rule must include the following, at a minimum: an explanation of the rule; an analysis of anticipated economic effects of the rule; a deadline for submitting comments; a tentative effective date; and the rule itself, as it would appear in the Code of Federal Regulations. If no adverse comments are received by the close of the comment period, the direct final rule becomes effective on the date specified. If any adverse comments are received by the close of the comment period, the direct final rule must be withdrawn. If APHIS chooses to proceed with rule-making, APHIS must issue a proposed rule. To ensure that the public receives notice of whether a direct final rule will become effective as indicated, APHIS publishes a brief notice after the comment period closes, either affirming the effective date or, if APHIS receives adverse comments, withdrawing the direct final rule.

RULEMAKING PROCESS IN APHIS

The following steps occur within APHIS before any rule is issued:

- A need is identified.
- A regulatory work plan is prepared, cleared within the Department of Agriculture, and designated as "significant" or "not significant" by OMB.
- A writer on the regulatory staff of APHIS receives the assignment and works with a technical expert from the relevant APHIS program area and others to develop the rule.
- The proposed rule is drafted and cleared within APHIS.
- The Office of General Counsel (OGC) of USDA reviews and clears the rule for legal sufficiency, and policy officials within the department also review and clear the rule. At a minimum, the policy officials include the Deputy Administrator for Animal Care or other program involved, the Administrator of APHIS, and the Under Secretary for Marketing and Regulatory Programs. If the proposed rule has been designated "significant" by OMB, it is also reviewed by other policy officials in the Department, including the Chief Economist of USDA, the Chief Information Officer, and the Secretary.
- The proposed rule is then also reviewed by OMB.
- After all clearances have been obtained, the proposed rule is signed by the Administrator or the Under Secretary and published in the *Federal Register*.

After the rule is published in the *Federal Register*, comments arrive. The comments undergo evaluation and, if changes are necessary, revisions will occur. A worksheet describing the number and nature of public comments received and the agency's planned response to them is submitted through the OMB for a designation (which may or may not be the same as the designation for the proposed rule). The final rule is drafted and cleared within APHIS. It is then reviewed and cleared by OGC, policy officials within USDA, and, if "significant," by OMB. The final rule is then signed and published in the *Federal Register* with a specified effective date.

TIME FRAMES

The time required for a given rulemaking varies, depending on the complexity of rule, the number and nature of comments received, the priority assigned by the agency (APHIS has an average of 150-200 actions in progress at any given time), and the designation assigned by OMB. Rules designated "significant" take longer than rules designated "not significant," at least partly because the clearance process within the department involves more policy officials and because OMB also reviews the document. OMB normally has 90 days to complete its review. Regulations designated "significant" may take several years to complete, from initiation of the regulatory workplan to publication of a final rule.

SUMMARY

For most requirements that the agency imposes on the public, the APA requires APHIS to conduct rulemaking. Although APA contains the basic requirements for rulemaking, including publication in the *Federal Register*, other laws and executive orders also apply to rulemaking. Among the various types of rulemaking documents, the most typical is a proposed rule followed by a final rule. Regulations are reviewed within APHIS, by the Office of General Counsel (USDA) and other policy officials, and, if designated "significant," by the OMB. Rules designated "significant" may take several years to complete.

ADDITIONAL INFORMATION

For more information about USDA-APHIS rulemaking, visit the APHIS website at www.aphis.usda.gov/index.html, or contact the Regulatory Analysis and Development Staff of APHIS at USDA-APHIS-RAD, 4700 River Road, Unit 118, Riverdale, MD 20737–1238, Phone: (301) 734-8682.

A Review and Comparison of Processes to Change Regulatory Guidelines: A European Perspective

Jonathan Richmond

The regulation of the use of animals for experimental and other scientific purposes and the determination of minimum required standards of animal care and accommodation across Europe are issues and processes noted more for their complexity and opaqueness than uniformity and transparency. Although each European member state devises and implements its own national legislation and standards of care and accommodation, these domestic provisions are informed, and in some cases determined, by supranational agreements and legislation at the level of the Council of Europe and the European Community.

The Council of Europe, which currently has 45 member states, was established in 1949 to protect human rights, to encourage both diversity and a common European identity, to seek problems to societal problems, and to consolidate democratic stability. Its focus is more societal and cultural than economical and political. A key Council instrument is the Convention—an agreement to impose common standards and practices that are binding only on the member states that choose to sign and ratify. Not only is ratification voluntary, but there are no legal penalties for not ratifying, or ratifying but not complying.

In contrast, the European Union (EU) is focused on economic and political union. Through directives and other statutory instruments, member countries are obliged to adopt common policies and approaches toward these ends. There are currently 15 member states, with a number of Eastern European "candidate countries" joining in the spring of 2004.

Although directives mandate WHAT has to be achieved, they allow each member state discretion as to HOW it is achieved. The European Commission monitors transposition and implementation, and noncompliance is dealt with through legal proceedings in the European courts. Matters are further complicated by the fact that the EU is itself a member of the Council of Europe and it seeks to represent EU countries within the Council of Europe on matters within the legal competency of the Union.

It is timely to reflect on how these processes work in practice, because current work on changing the Council of Europe Convention and the European Community Directive provides useful insights into how changes are made to the content of these European instruments.

COUNCIL OF EUROPE CONVENTION ETS 123 (1986)

Council of Europe Convention ETS 123, which dates from 1986, makes provision for the protection of animals produced and used for experimental and other scientific purposes, including both fundamental and applied research. Most members of the European Union, and the Commission itself, have signed and ratified this Convention. Appendix A of the Convention sets minimum provisions for the housing and care of animals. Strictly speaking, those who have ratified the Convention are required only to "take note" of these provisions.

The Council has been working for several years now on revising Appendix A. Initial hopes in some quarters that the revision would produce evidence-based optimum standards of animal care and accommodation quickly proved unrealistic. Even for the common laboratory species, there is a dearth of evidence-based material on laboratory animal care and accommodation (and the Council has neither the time nor the resources to generate evidence), there are numerous opinions about contemporary best practice, but there is no means of demonstrating optimum provision.

In an attempt to resolve this problem, the Council of Europe established a series of technical expert working groups that were supported by a Secretariat, a steering group including representatives of the parties to the Convention, and various observers, including the Institute for Laboratory Animal Research, the US Department of Agriculture, and the Canadian Council on Animal Care, and a separate drafting group. The working groups were charged with the task of producing both general and animal-specific guidance on care and accommodation. The expert groups are broadly based and consist of experts largely drawn from nongovernmental organizations but excluding the national competent authorities.

The technical work is nearing completion. Both general guidance and species-specific provisions are being drafted (backed up by separate detailed appendices that set out the information that was considered when the revised specifications were determined). Once the technical content has been finalized, a multilateral meeting of the parties to the Convention will be convened to adopt or reject the proposals.

Two key considerations are evident in the outputs to date: (1) The quality and complexity of the space provided is as important as the area or volume of space; and (2) pair and group housing of social species will be considered to be the norm. Most importantly, to encourage and facilitate diversity and innovation, "performance-based" standards are being sought whenever practical. The Council of Europe is placing drafts of "completed" documents on its website. This process has been expensive, lengthy, and complex—and it is not finished yet.

Although the status of the relevant Appendix to the Convention is likely to be as before (i.e., something the Parties should take note of), the fact that the European Community is now a Party to the Convention is of significance. Specifically, the Community will be obliged to ensure that similar standards are incorporated into the relevant European Directive, and member states may then be obliged to conform.

EUROPEAN UNION DIRECTIVE 86/609/EEC (1986)

The statutory instrument EU Directive 86/609/EEC makes provision for the harmonization of laws for the protection of animals produced and used for experimental and other scientific purposes. In scope, it covers safety testing; work aimed at preventing, detecting, and controlling disease; the assessment, detection, and modification of physiological functions; and protection of humans and the natural environment. However, unlike the Convention, it probably does not cover fundamental research (including drug discovery), forensic enquiries, or education and training.

The Directive includes an "informative Annex" on standards of care and accommodation that the member states shall "pay regard to." Furthermore "animal welfare" is not a Commission competence strictly speaking, and the Directive permits member states to "adopt stricter measures." Work on reviewing and revising the Directive started recently. The early background work is being undertaken by four Technical Expert Working Groups, which are again weighted in favor of nongovernmental organizations rather than national competent authorities. Even at this stage, it is clear that the 3Rs of replacement, reduction, and refinement; a system of harm/benefit assessment; and a requirement for ethical review processes will underpin a revised Directive.

Another EU Directive that has an impact on animal use is the establishment of criteria to which products (e.g., chemicals, pharmaceuticals, and medical devices) must conform before they can be marketed or used within member states. Ensuring that these criteria make best provision for a free market, consumer choice and safety, and animal welfare is a real challenge. Furthermore, the need for the EU to ensure that proposed test methods that replace, reduce, or refine animal use are scientifically valid before changing current test requirements is also a challenge. The role of the European Centre for the Validation of Alternative Methods and the Interagency Coordinating Committee on the Validation of Alternative Methods will also be considered.

HARMONIZATION AND DIVERSITY

Within Europe, there might be one Directive and one Convention to protect animals used for scientific purposes; but even within the EU, there are nevertheless 15 ways of doing it. Within the EU, the various national systems give a first impression of diversity, rather than harmony. Nevertheless, although the details vary, they have many elements in common; and much of the meeting presentation will compare and contrast key elements of several national systems. Their commonalities include regulation impacts on the place, the program of work, the personnel involved, and the training requirements for key staff. Authorization may be invested centrally, regionally, or even locally; but oversight by inspection is a common provision. Ethical review processes have become the norm, with different functions being discharged at local, regional, and national levels.

Hopefully, and mindful of the reasons for there being an EU, the revision of the Directive will focus on what needs to be regulated, rather than on what can be regulated; and it will produce a proportionate system not overly endowed with bureaucracy. Although all member states welcome the flexibility of approach inherent in Directives, all are agreed that the outputs must be harmonized. Yet we live in interesting times, and it will be interesting to see what priority of the political agenda these issues are assigned when a newly elected European Parliament convenes next year. This Parliament will be composed of representatives of current member states and candidate countries.

Japanese Regulations on Animal Experiments: Current Status and Perspectives

Naoko Kagiyama and Tatsuji Nomura

In terms of ethics in animal experimentation, advanced countries have adopted the 3R principles of humane experimental technique first espoused by Russell and Burch in 1959, namely, replacement, reduction, and refinement. Twenty-six years later, in 1985, the 3Rs were translated into 11 basic principles by the Council of International Organizations for Medical Sciences (CIOMS). These items have become international principles that govern animal experimentation.

INTERNATIONAL COMPARISON OF REGULATIONS ON ANIMAL EXPERIMENTATION

Europe

The year 1986 was important in terms of the following: (1) the Council of Europe concluded the convention for the protection of vertebrate animals used for experimental and other scientific purposes; (2) the European Union formulated directives on the approximation of laws, regulations, and administrative provisions to rectify disparities in welfare policies among member states; and (3) in the United Kingdom, the Cruelty to Animals Act of 1876 was amended and its title changed to the Animals (Scientific Procedures) Act (Figure 1, left). The regulatory agency in the United Kingdom, the Home Office, issues three licenses—for the project, the personnel, and the premises where the animal experiment is to be

CoE ETS123 & EU Directive 86/609

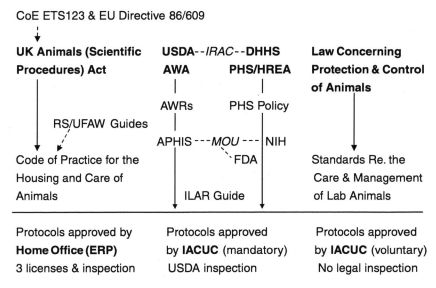

FIGURE 1 Comparison between regulations on animal experiments in the United Kingdom, the United States, and Japan. USDA, US Department of Agriculture; IRAC, Interagency Research Animal Committee; DHHS, Department of Health and Human Services; AWA, Animal Welfare Act; PHS, Public Health Service; HREA, Health Research Extension Act; AWR, Animal Welfare Regulations; UFAW, Universities Federation for Animal Welfare; APHIS, Animal and Plant Health Inspection Service; MOU, Memorandum of Understanding; NIH, National Institutes of Health; FDA, Food and Drug Administration; ILAR, Institute for Laboratory Animal Research; IACUC, institutional animal care and use committee.

conducted. All three licenses must be obtained before starting the animal experiment. Thus, the UK regulatory authority directly controls animal experimentation.

United States

In the United States, the Animal Welfare Act was enacted in 1966. The amendment in 1985 required research facilities to appoint an institutional animal care and use committee (IACUC) as well as to ensure the qualifications of personnel involved in animal experiments (Figure 1, center). Animal experiments can be performed based on a review and approval of the IACUC and the final approval of the institutional official. The regulatory agency, the US Department of Agriculture (USDA), conducts unannounced inspections of facilities every year.

Japan

In Japan, animal experimentation is also regulated by laws. The types of regulations in Japan (Figure 1, right) resemble the US system, which holds each institution responsible for self-regulation. However, the designation of an equivalent of the IACUC, registration, and legal inspections of laboratory animal facilities are not stipulated in the law.

LEGAL SYSTEMS IN JAPAN

The detailed legal system in Japan is described in Figure 2 from an historical viewpoint. The Law Concerning the Protection and Control of Animals enacted in 1973 was amended in 1999 and given the new title of the Law for the Humane Treatment and Management of Animals. This law protects all species of animals from cruelty (Investigative Committee 2001).

The law emphasizes respect for life, companionship with animals, and well-being of animals. It specifies the responsibility of the owner of the animal, and calls for the alleviation of pain and distress as well as the humane death of animals used for scientific purposes. Based on the law, the Standards Relating to the Care and Management of Experimental

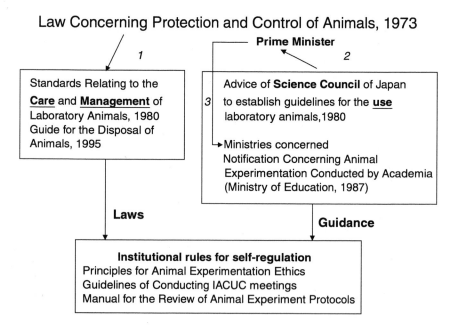

FIGURE 2 Regulations related to animal experiments in Japan.

Animals were specified in 1980 (Figure 2, left). The Standards cover the *care* and *management* of laboratory animals, but not the *use* of animals (Investigative Committee 1980). The same year that the Standards were announced, the Science Council of Japan advised the Prime Minister to prepare administrative guidance for the *use* of animals for scientific purposes (Science Council of Japan 1981).

In 1987, in accordance with the advice of the Science Council of Japan, the Ministry of Education notified universities and other institutions to establish voluntary guidelines on animal experimentation (Ministry of Education 1987) (Figure 2, right). As a result of the notification, universities and even private research institutes formulated their own principles, guidelines, manuals, and other materials in accordance with the laws and with administrative guidance. Similarly, the scientific associations concerned have compiled guidelines for individual research fields to encourage members to balance science and animal welfare (JALAS 1987). Thus, animal experiments in Japan are regulated by a combination of legal and scientific developments.

CURRENT STATUS

Current regulations regarding animal experimentation in Japan are summarized in Table 1. The three categories include the following: laws consisting of the law, the standards, and the guide; administrative guidance issued by the Head of Science and International Affairs Bureau through their bulletin, which includes the notification; and voluntary guidelines formulated by individual scientific associations. Laboratory animal scientists in Japan have observed all three categories of regulations equally without the force of law but with a moral sense. The law, the standards, and the guide in the first category consist of 31, 10, and 4

TABLE 1 List of Regulations Regarding Animal Experiments in Japan

1. **LAWS**
   Law for the Humane Treatment and Management of Animals (Ministry of the Environment)
   — Standards Relating to the Care and Management, etc. of Experimental Animals
   — Guide for the Disposal of Animals
2. **ADMINISTRATIVE GUIDANCE**
   Notification Concerning Animal Experimentation Conducted by Universities, etc. (Ministry of Education)
3. **VOLUNTARY GUIDELINES**
   More than 12 guidelines on animal experiments formulated by individual scientific associations

TABLE 2 Characteristics of Regulations on Animal Experiments in Japan

- Laws specify the responsibility of the owner of the animal
- Laws call for 3Rs emphasizing the alleviation of pain and distress as well as humane death of animals used for scientific purposes
- Administrative guidance encourages the ethical use of animals
- Self-regulation system similar to the US and Canada
- Recommendation for designating IACUC
- Exemption from legal registration/inspection

articles, respectively. Each has an associated explanatory handbook with 302, 115, and 68 pages of texts, respectively. The handbook for the Standards was edited by seven laboratory animal scientists, one medical doctor, and one representative of the Japan Animal Welfare Society to reflect the opinions of animal advocates.

In Table 2, the characteristics of Japanese regulations on animal experiments are listed. Our ethical principles, known as the 3Rs, are the same as in Western countries. The law emphasizes refinement that focuses on the alleviation of pain and distress as well as the humane death of animals.

The law adopts the self-regulation system for animal experimentation, and the notification only *recommends* designation of an IACUC (Ministry of Education 1987). Nevertheless, according to survey results, almost all medical schools and pharmaceutical companies as well as about one-third of breeders have established IACUCs even though the law does not mandate it. Laboratory animal and livestock facilities are exempted from registration and legal inspection. Instead, the Japanese Association for Laboratory Animal Science voluntarily conducts surveys every 3 years on the number of animals used for scientific purposes (CLACU 2003). It is therefore evident that laboratory animal scientists in Japan recognize the importance of replacement of live animals with insentient materials and reduction in the number of animals involved.

Although our regulations may appear somewhat lenient and ambiguous from the Western viewpoint (Nomura 1995) (a feeling that has sometimes annoyed Japanese scientists when collaborating with Western colleagues), the authors believe that certain religious implications may underlie animal experimentation ethics and the structure of regulations in individual countries, as described below.

RELIGIOUS IMPLICATIONS

A Thai venerable who graduated from medical school made a presentation about the philosophy of *Karma* in Buddhism at the 2003 annual meeting of the American Association of Laboratory Animal Sciences. Researchers in Buddhist countries regard animals as existing on the same level as humans and generally treat animals based on the philosophy of *Karma*. Buddha rewards or punishes people based on their deeds *(Karma)*. The logic of reincarnation is called *samsara*, the endless round of rebirth and redeath based on the impersonal judge of the Natural World, on which the destiny of humans depends.

Researchers in historically Christian countries, by comparison, appear to handle laboratory animals from the standpoint of the Lord of Creation. Researchers in Western countries give the impression of conducting animal experiments based on phylogenic domination of the animal. Therefore, they often cite the expression "humane care and responsible use."

PERSPECTIVE OF ANIMAL EXPERIMENTS IN JAPAN

In Japan, we seek to combine Buddhist and Christian assumptions and to reach a point where humans should take responsibility for laboratory animals so that they can accomplish good *Karma* while using the animals for scientific purposes. If animals are suffering from infectious diseases, for example, they will not be able to provide reliable experimental data as their good *Karma*. Thus, microbiological control of the animal environment should be the responsibility as well as good *Karma* for people engaged in animal experimentation. Investigators in Japan may achieve the 3Rs principle in practice without any strict regulations because of the fear of *samsara*. However, the need to collaborate with Western colleagues requires compromise with the Western system. For this reason, we have been discussing appropriate strategies.

There is current disagreement over whether we should aim toward more stringent regulations, similar to European countries, or continue the current self-regulation system as in the United States and Canada. To ensure a convincing self-regulation system, we will need to clarify the responsibility of persons who engage in animal experiments, define the role of the IACUC, and implement animal welfare practices compatible with scientific needs, as shown in Figure 3. As mentioned above, none of these elements are strictly stipulated by the law but are regulated by administrative guidance and voluntary guidelines to encourage flexible animal research. It is now time for us to consider a certain mechanism to defend animal research as well as a validation system for self-regulation to reach a social consensus on the necessity of animal experimentation.

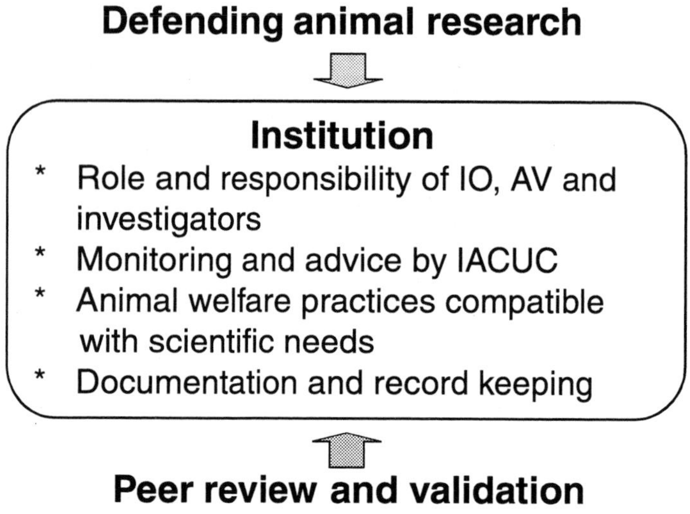

Defending animal research

Institution

* Role and responsibility of IO, AV and investigators
* Monitoring and advice by IACUC
* Animal welfare practices compatible with scientific needs
* Documentation and record keeping

Peer review and validation

FIGURE 3 Reinforcement of the self-regulation system within animal research facilities in Japan.

REFERENCES

CLACU [Committee for Laboratory Animal Care and Use, Japanese Association for Laboratory Animal Science]. 2003. The number of live animals used in experiments in 2001 [text in Japanese, summary in English]. Exp Anim 52:143-158.

Investigative Committee [Investigative Committee on the Law for the Humane Treatment and Management of Animals]. 2001. Explanatory Handbook on the Law [text in Japanese]. Tokyo: Seirin Shoin.

Investigative Committee [Investigative Committee on the Standards Relating to the Care and Management of Experimental Animals]. 1980. Explanatory Handbook on the Standards [in Japanese]. Tokyo: Gyousei.

JALAS [Japanese Association for Laboratory Animal Science]. 1987. Guidelines on Animal Experimentation [in Japanese]. Exp Anim 36:285-288.

Ministry of Education [Ministry of Education, Science and Culture, Science and International Affairs Bureau]. 1987. Notification on Animal Experimentation in Universities [in Japanese].

Nomura, T. 1995. Laboratory animal care policies and regulations in Japan. ILAR J 37:60-61.

Science Council of Japan. 1981. Recommendation for the establishment of animal experiment guideline [in Japanese]. Exp Anim 30:173-178.

Process for Change—Development and Implementation of Standards for Animal Care and Use in Canada

Clément Gauthier

The Canadian Council on Animal Care (CCAC) is the national organization that has set and overseen the implementation of standards for the care and use of animals in science since 1968. It is a peer review organization involving close to 2000 scientists, veterinarians, animal care technicians, students, community representatives, and representatives of the animal welfare movement through its programs of assessment, guidelines development, education, and training. The CCAC pioneered the institutional animal care committee (ACC) as the local keystone of its decentralized ethical review and oversight system. For a detailed description, consult the following documents of the CCAC in the order indicated: *The Assessment Program of the CCAC*, 2000; *Terms of Reference for Animal Care Committees*, 2000; CCAC *Guidelines on Animal Use Protocol Review*, 1997; and *Categories of Invasiveness in Animal Experiments*, 1991. All of these documents are available on the CCAC web site [http://www.ccac.ca]. ACCs are now part of most legislated and voluntary frameworks for the regulation/oversight of animals used for scientific purposes worldwide.

Although compliance with CCAC standards is voluntary for government and industry, it is mandated through federal spending power for academic institutions. Over the past decade, provincial and federal governments have increasingly recognized the CCAC as a quasi-regulatory system. CCAC standards are now referenced in the regulations to provincial laws relating to the use of animals for scientific purposes (CCAC

2003; Russell and Burch 1959) and in the policies of relevant federal departments and agencies.

The underlying ethical basis of all CCAC guidelines and policies requires adherence to the 3Rs (reduction, replacement, and refinement) of Russell and Burch (1959). Maximizing animal well-being and minimizing pain and distress are the ethical drivers for the development of refinement measures. Housing conditions and environmental controls must be refined with the primary objective of meeting the social and behavioral needs of animals and of maximizing animal well-being. To minimize pain and distress, investigators frequently adopt practices believed to improve animal welfare based on anecdotal evidence. Although there is no commonly understood definition for the term "best practice," per se, there is agreement on its content criteria and consensus on the fact that it must be evaluated by peers. The peer-based approach underlying the CCAC guidelines development process and the use of evidence-based learning loops in the evolution of best practices to implement these guidelines are two essential pillars of international harmonization of standards for the care and use of animals in science.

Compliance with CCAC standards is ensured at the local level through ACC peer-based review of protocols and mandatory site visits of facilities. National quality assurance is provided through CCAC's external, peer-based assessment visits, which include the assessment of the functioning of the ACC. At the June 2001 first International Symposium on Regulatory Testing and Animal Welfare organized by the CCAC in collaboration with the International Council for Laboratory Animal Sciences, the Breakout Group on Best Practices for Animal Care Committees and Animal Use Oversight concluded that future progress requires encouraging diversity of frameworks as a source of continuous improvement, and the networking of ACCs to identify, encourage, and share best practices (ILAR 2002).

The institutional ACC plays a central role as the third pillar of international harmonization, for the following reasons:

1. It is representative of the scientific culture and moral values of home countries;

2. It facilitates communications and empowers informed decision making at the local level;

3. It is already integrated as an accountable keystone of most national oversight and regulatory systems worldwide; and

4. It provides each country with enhanced ability to influence international harmonization of best practices for animal care and use in science.

REFERENCES

CCAC [Canadian Council on Animal Care]. 2003. Legislation in Canada related to experimental animals. In: Experimental Animal User Training Core Topics Modules, Module 1: Guidelines, Legislation and Regulations. Electronic document (http://www.ccac.ca/english/educat/Module01E/module01-11.html).

ILAR [Institute for Laboratory Animal Research]. 2002. Proceedings of the ICLAS/CCAC International Symposium held in Québec City, Canada, June 21-23, 2001. ILAR J 43:S1-S136. Available at http://dels.nas.edu/ilar/jour_online.

Russell, W.M.S., and Burch, R. 1992 [1959]. Principles of Humane Experimental Technique. England: Universities Federation for Animal Welfare, Potters Bar, Herts, UK.

Building Credible Science from Quality, Animal-based Information

Paul Gilman

The research and development arm of the US Environmental Protection Agency (EPA) employs 1,950 employees at 13 laboratories and research facilities across the nation. Its $700 million budget for fiscal year 2003 included a $100 million extramural research grant program. The focus of those people, facilities, and programs is the generation of credible, relevant, and timely research results—and technical support—that inform EPA's policies, decision making, and promulgation of regulations. Setting those regulations, making those policies, and accomplishing those decisions with sound science requires the Office of Research and Development (ORD) to fulfill the following responsibilities:

- Production of relevant, high-quality, cutting-edge research results in human health, ecology, pollution control, and prevention investigations and in economics and decision sciences;
- Characterization of scientific findings properly; and
- Use of appropriate tools and approaches for the performance of science in the decision process.

High-priority areas for ORD include research in human health, particulate matter, drinking water, clean water, global change, endocrine disruptors, ecological risk, pollution prevention, and homeland security. Each of these important areas of investigation may include the use of animal models.

In its role as a leader in the environmental and human health protection research communities, ORD upholds three fundamental tenets in the use of animals in its research:

- Animals represent a limited resource and are precious research tools;
- Quality and validity of data collected from animals must be assured; and
- Resultant research must have the full confidence of and acceptance by the scientific, regulated, and decision-making communities, and the public.

Understanding the specific needs for and value of animal use is balanced with associated responsibilities in conducting such research. ORD adheres to

- Stringent review to assure responsible use;
- Annual review and accreditation by the Association for Assessment and Accreditation of Laboratory Animal Care;
- Close scrutiny of animal protocols; and
- Continuous attention to the refinement of methodologies that reduce the number of or eliminate the use of animals.

Session 2

Assessment of Animal Housing Needs in the Research Setting— Peer-reviewed Literature Approach

Disruption of Laboratory Experiments Due to Leaching of Bisphenol A from Polycarbonate Cages and Bottles and Uncontrolled Variability in Components of Animal Feed

Frederick S. vom Saal, Catherine A. Richter,
Rachel R. Ruhlen, Susan C. Nagel, and Wade V. Welshons

Mammalian embryonic development is epigenetic in that hormonal signals not only control the timing of gene expression but also set the activity of genes and thus the functioning of organs and homeostatic systems for the remainder of life. Variation in endogenous hormones (e.g., estradiol and testosterone), which regulate the development of organs (vom Saal 1989), or disruption of the activity of these hormones during development by chemicals can lead to permanent changes in organ structure and function. Adult exposure to endocrine-disrupting chemicals can lead to transient changes in organ function that can disrupt experiments.

Polycarbonate cages and water bottles are manufactured by polymerizing the chemical bisphenol A, which was initially considered for use as an estrogenic drug before being used to manufacture polycarbonate in the 1950s (Dodds and Lawson 1936). More than 50 published studies have shown effects of developmental as well as adult exposure to bisphenol A on a wide variety of traits in mollusks, insects, fish, frogs, rats, and mice. Polycarbonate cages have been commonly used to house rodents and aquatic animals in laboratory experiments. What was not appreciated by scientists using these cages until recently is that after repeated washings the rate of leaching of bisphenol A increases dramatically and can reach levels that can alter traits in animals. Howdeshell and coworkers reported that a small but detectable amount of bisphenol A leached out of new polycarbonate animal cages into water at room temperature, and the rate of leaching was more than 1000 times greater ($> 300 \ \mu g/L$) in old, visibly worn (scratched and discolored) polycarbonate cages (Howdeshell and others 2003).

Hunt and colleagues reported that an adverse effect of exposure to very low doses of bisphenol A leaching from polycarbonate animal cages and water bottles is profound disruption of chromosomes during meiosis in oocytes in female mice. Specifically, there was a dramatic increase in the incidence of abnormal alignment of chromosomes during the first meiotic division in oocytes, which was caused by the leaching of bisphenol A from the polycarbonate cages washed with a harsh detergent (Hunt and others 2003; Koehler and others 2003). Abnormal alignment of chromosomes results in aneuploidy, or abnormal numbers of chromosomes in oocytes, which can lead to abnormal development such as occurs in Down's syndrome. These authors thus refer to bisphenol A as a "potent meiotic aneugen." Aneuploidy is thought to be a major cause of embryonic mortality in humans. Hunt and coworkers reported that severe oocyte chromosome abnormalities increased in peripubertal female mice in the following proportions: from a baseline frequency of 1.8% in control animals (not housed in damaged cages) to 20% due to housing females in damaged polycarbonate cages; 30% due to the use of damaged polycarbonate water bottles; and 41% due to combined use of both damaged cages and water bottles. In a subsequent experiment, the researchers intentionally accelerated the normal aging process associated with repeated washing of polycarbonate cages and water bottles by washing them different numbers of times in a harsh detergent. The polycarbonate water bottles were found by gas chromatography/mass spectrometry analysis to release between 100 (mild damage) and 260 µg/liter (severe damage) of free bisphenol A into water placed into the bottles, resulting in daily exposure of the female mice ranging between 15 and 72 µg/kg. When peripubertal female mice housed in undamaged new cages were fed bisphenol A once daily at the very low doses of 20, 40, and 100 µg/kg to simulate exposure within the range released by the polycarbonate, there was a significant dose-related increase in the incidence of chromosomal damage beginning even at the lowest dose.

Based on studies in which bisphenol A was found to have limited binding to the plasma proteins that serve as a barrier to the movement of estrogen from blood into tissues (Nagel and others 1997), we had predicted that doses of bisphenol A as low as 20 µg/kg/day would disrupt development in mice. This dose is below the predicted "safe" or reference dose for human exposure of 50 µg/kg/day, which was calculated based on old studies that examined only very high doses of bisphenol A, when the lowest dose administered (50 mg/kg/day) had resulted in adverse effects (IRIS 2002).

Taken together, the large number of independent findings concerning adverse effects of very low doses of bisphenol A suggest that the use of polycarbonate to manufacture animal cages and water bottles can alter the results of laboratory animal research. In fact, due to greater resistance to heat and alkaline detergents, many facilities have switched from polycarbonate to polysulfone animal cages. The ether bond in the polysulfone co-polymer is more resistant to heat and alkaline detergents relative to the ester bond in polycarbonate. We have always used polypropylene cages and glass water bottles, because these cages do not leach biologically active amounts of estrogenic chemicals (Howdeshell and others 2003).

Some components of feed used in laboratory experiments (e.g., phytoestrogens and mycotoxins) have hormonal activity that can interfere with experiments involving outcomes that are sensitive to these hormone-mimicking chemicals. There is wide variation in phytoestrogen content in different types of commercial rodent feed. Both the amount of phytoestrogens and metabolizable energy in different feeds were sources of phenotypic variation (specifically body weight, uterine growth, and age at vaginal opening) in prepubertal CD-1 female mice (Thigpen and others 2003). Thigpen and colleagues selected one soy-based commercial feed (Purina 5002) and examined the consequences of using five batches of this diet with different mill dates. They first measured the amounts of the soy phytoestrogens genistein and daidzein in the five different batches, which ranged from 159 to 431 μg/g. It is well known that one of the effects of the estrogenic drug diethylstilbestrol (DES) is to accelerate vaginal opening, and although a 4-ppb dose of DES accelerated vaginal opening in female CD-1 mice fed the batch of feed with 159 μg/g of genistein and daidzein, there was no accelerating effect of DES in females being fed the batch of feed with 431 μg/g of genistein and daidzein. Administration of even this potent estrogenic drug could not further accelerate this process (Thigpen and others 2003). A similar finding had previously been reported for prepubertal female rats fed different batches of feed produced by another feed manufacturer (Boettger-Tong and others 1998). Together, these studies reveal that the issue of variation in phytoestrogen content in batches of feeds is one that is a general problem and not just restricted to Purina 5002 feed, because any closed-formula diet can contain variable amounts of phytoestrogens due to the source as well as the amount of soy isoflavones.

It is important to emphasize that the isoflavones genistein and daidzein are only two of many naturally occurring compounds that could be sources of estrogenic activity in feed, and even casein-based feeds show variation in total estrogenic activity (Thigpen and others 2003). For example, we have found significant variation in estrogenic activities in different batches of casein-based feeds, and none of these estrogens were genistein or daidzein (unpublished observation). Simply screening for these two isoflavones will thus not guarantee the lack of variability in other potential endocrine-disrupting contaminants in a feed. Feed manufacturers need to develop new approaches to reduce variability in endocrine-disrupting activity in different batches of feeds to attain levels that will not disrupt research results.

We have also found that there are components of some batches of commercial mouse feeds, such as soy-based Purina 5002 certified diet, that, relative to other feeds (Purina 5008 soy-based pregnancy diet), dramatically increase endogenous estradiol in CD-1 mouse fetuses (unpublished observation). This increase is associated during later life in both males and females fed Purina 5002 throughout life with an increase in postnatal rate of growth, accelerated onset of puberty in females, and an increase in the amount of abdominal fat. Male mice fed Purina 5002 diet also evidenced differences in reproductive organs, such as an increase in prostate size and a decrease in daily sperm production, relative to males whose mothers were fed Purina 5008 during pregnancy and lactation, followed by soy-based Purina 5001 after weaning. An interesting additional finding is that oral administration of DES to pregnant mice of a low dose ($0.1 \ \mu g/kg/day$) and a high dose ($50 \ \mu g/kg/day$) resulted in a dose-related decrease in daily sperm production in adult male offspring on the Purina 5008/5001 regimen, whereas males from the Purina 5002 regimen showed no effect of DES, even at the high dose (unpublished observation). Most investigators are aware of the marked effects that different types of feed can have on the phenotype of their animals. Of great concern, however, is that variability between batches of some feeds is a potential source of uncontrolled variability in research results.

Another variable of concern in laboratory studies is water quality. Copper pipes are used inside the building that houses our mice, and water is provided to the mice in glass bottles. We purify the water by ion exchange and a series of carbon filters. These measures are important to remove contaminants such as phthalates, which can enter water from PVC water pipes and herbicides. These potential contaminants, as well as other solutes that can affect an animal's physiology, are an issue particularly in agricultural areas.

ACKNOWLEDGMENTS

Support was provided during the preparation of this paper from grants to F.vS. from the National Institute of Environmental Health Sciences (NIEHS) (ES11283), C.A.R. from NIEHS (ES-11549), S.C.N. from the National Institute of Diabetes and Digestive and Kidney Diseases (DK60567), and W.V.W. from the National Cancer Institute (CA50354) and the University of Missouri (VMFC0018).

REFERENCES

Boettger-Tong, H., Murthy, L., Chiappetta, C., Kirkland, J.L., Goodwin, B., Adlercreutz, H., Stancel, G.M., Makela, S. 1998. A case of a laboratory animal feed with high estrogenic activity and its impact on in vivo responses to exogenously administered estrogens. Environ Health Perspect 106:369-373.

Dodds, E.C., and Lawson, W. 1936. Synthetic oestrogenic agents without the phenanthrene nucleus. Nature 137:996.

Howdeshell, K.L., Peterman, P.H., Judy, B.M., Taylor, J.A., Orazio, C.E., Ruhlen, R.L., vom Saal, F.S., Welshons, W.V. 2003. Bisphenol A is released from used polycarbonate animal cages into water at room temperature. Environ Health Perspect 111:1180-1187.

Hunt, P.A., Koehler, K.E., Susiarjo, M., Hodges, C.A., Hagan, A., Voigt, R.C., Thomas, S., Thomas, B.F., and Hassold, T.J. 2003. Bisphenol A causes meiotic aneuploidy in the female mouse. Current Biol 13:546-553.

IRIS, Bisphenol, A. (CASRN 80-05-7), US-EPA Integrated Risk Information System (IRIS). Substance file. http://www.epa.gov/iris/subst/0356.htm. Accessed January 2002.

Koehler, K.E., Voigt, R.C., Thomas, S., Lamb, B., Urban, C., Hassold, T.J., Hunt, P.A. 2003. When disaster strikes: Rethinking caging materials. Lab Anim 32:32-35.

Nagel, S.C., vom Saal, F.S., Thayer, K.A., Dhar, M.G., Boechler, M., Welshons, W.V. 1997. Relative binding affinity-serum modified access (RBA-SMA) assay predicts the relative in vivo bioactivity of the xenoestrogens bisphenol A and octylphenol. Environ Health Perspect 105:70-76.

Thigpen, J.E., Haseman, J.K., Saunders, H.E., Setchell, K.D.R., Grant, M.G., Forsythe, D.B. 2003. Dietary phytoestrogens accelerate the time of vaginal opening in immature CD-1 mice. Comp Med 53:477-485.

vom Saal, F.S. 1989. Sexual differentiation in litter-bearing mammals: influence of sex of adjacent fetuses in utero. J Anim Sci 67:1824-1840.

Assessment of Animal Housing Needs in the Research Setting Using a Peer-reviewed Literature Approach: Dogs and Cats

Graham Moore

INTRODUCTION

A previous presentation (De Leeuw 2004) provided an outline of the Council of Europe (CoE) and the background of its process to revise Appendix A (guidelines for accommodation and care) of Convention ETS 123 (European Convention for the Protection of Vertebrate Animals used for Experimental and Other Scientific Purposes) of 1986. In this presentation, I will describe how the Council of Europe's requirements have been applied to the preparation of draft guidance on the accommodation and care of dogs and cats.

COUNCIL OF EUROPE PROCESS FOR THE REVISION OF APPENDIX A

In 1997, the CoE adopted a **resolution on accommodation and care** of vertebrate animals used for experimental and other scientific purposes (Council of Europe 1997). It had been generally agreed by the CoE that the existing "Guidelines for accommodation and care of animals" presented in Appendix A of Convention ETS 123 had proved very useful and had been applied widely within Europe. However, it was also acknowledged that scientific knowledge and experience had progressed since 1986 and the entry into force of the Convention, such that a review of the guidelines was necessary.

The resolution stipulated that the new proposals should be divided into General and Species-Specific recommendations and also indicated key areas to which attention should be given. It further identified the way in which guidance should be prepared. **Expert Groups,** with representation from nominees of observer nongovernmental organizations of the Council of Europe but not of the national authorities, were to be set up to prepare proposals on the main groups of species covered by the Convention. These proposals would then be submitted to a **Working Party** for comment, amendment, and endorsement. Membership of the Working Party comprised representatives of the national authorities of the CoE member states, together with observers from a wide range of concerned nongovernmental organizations. A **Drafting Group** assisted in the work of the Expert Groups and of the Working Party. Once the Working Party agreed on all proposals from the Expert Groups, they would be received by a CoE **Multilateral Consultation** for any further discussion and approval, before being submitted to the **Committee of Ministers** for final approval.

It is important to note that the **status of Appendix A is "guidance" and the guidelines are not mandatory.** However, it was generally considered by the Expert Groups that these should be regarded as minimum requirements.

CURRENT STATUS OF THE REVISION

Initially, only four Expert Groups had been established, on (1) Rodents and Rabbits, (2) Dogs and Cats, (3) Nonhuman Primates, and (4) Pigs and Minipigs. These four groups were given the task of preparing Species-Specific proposals, with the General Part of the new proposals, including provisions common to all species covered, being drafted with input from all four groups. The Working Party later decided to add additional species covered by the Convention to the list of those already to be covered by the revision; thus, the number of groups grew from four to eight. Furthermore, the Pigs and Minipigs group was expanded to cover all farm animal species, and ferrets were added to the Dogs and Cats group.

Currently, the General Section and Species-Specific proposals for Rodents, Rabbits, Dogs, Cats, and Ferrets have been finalized by the Working Party. Those for other species have not yet been finalized, although those for Nonhuman Primates, Birds, and Amphibians are at a very advanced stage.

MEMBERSHIP OF EXPERT GROUP ON DOGS AND CATS AND *MODUS OPERANDI*

As with all of the Expert Groups, there was a broad-based representation drawn from the observer nongovernmental organizations of the CoE. The membership of the Expert Group on Dogs and Cats comprised one representative from each of the following: the Eurogroup for Animal Welfare, the European Federation of Pharmaceutical Industries and Associations (EFPIA), the Federation of Laboratory Animal Breeders Associations (FELABA), the Federation of Veterinarians in Europe (FVE), and the International Society for Applied Ethology (ISAE). Meetings were coordinated and chaired by a representative of the Federation of European Laboratory Animal Science Associations (FELASA). A second ISAE representative was subsequently added because more input on cat ethology was deemed necessary. This membership was thought to provide a broad spread of expertise and opinion, which would result in the formulation of an expert view on minimum standards for these species.

It was decided at an early stage that the group would work primarily on the basis of face-to-face meetings (held in London or Brussels), with e-mail communication between meetings. Additional input would be sought as necessary from within represented organizations or from other experts. The Coordinator of the Group, together with one or more members, attended all meetings of the Working Party in Strasbourg to present the Group's proposals, discuss their content and answer questions, and refer matters back to the Group as appropriate.

BASIS FOR DOG AND CAT RECOMMENDATIONS

The CoE stipulated the provision of proposals for a General Section and for Species-Specific Sections (called Part A). It also requested a supporting explanatory and referenced text (Part B) for each of the sections.

Groups were directed to pay special attention to enrichment of the environment, particularly in relation to social interactions, activity-related use of the space, and provision of appropriate stimuli and materials. Proposals were to be based on science-based information when it was available, and otherwise on practical experience and good or "best" practice. Where appropriate, Expert Groups were given the task of identifying areas in which additional research would be desirable.

The Expert Group on Dogs and Cats considered these areas and paid attention to existing guidance documents, such as the current Appendix A, UK Home Office guidance (1989, 1995), and the ILAR *Guide for the Care and Use of Laboratory Animals* (NRC 1996). Some significant variations were found in these recommendations, particularly those for space requirements and for some environmental parameters. It was therefore decided that where robust science-based information was not available, proposals would be based primarily on an examination of the animals' physiological and ethological needs, taking into account the current views on good/best practice and the inevitable constraints of a research environment.

Species-Specific Section—Subject Headings

The species-specific sections for dogs and cats covered the following:

Preamble
Introduction
The environment and its control
    Ventilation
    Temperature
    Humidity
    Lighting
    Noise
    Alarm systems
Health
Housing and enrichment
    Housing
    Early socialization with conspecifics and humans
    Enrichment
    Animal enclosures
        Outside runs (dogs only)
        Dimensions
        Flooring
Feeding
Watering
Substrate, litter, bedding, and nesting material
Cleaning
Handling (cats only)
Humane killing
Records
Identification

For subject headings that were common to all species-specific sections, the Group decided no proposals were necessary other than those already contained in the General Section. These headings are shown in italics in the list above. Additional headings in the General Section were Definitions; Physical Facilities; Education and Training; Care (incorporating

Health, Capture from the wild, and Transport); and Quarantine, acclimatization, and isolation.

Space does not permit detailing all of these recommendations; however, some examples drawn from the proposals on housing and enrichment and on dimensions and flooring provide an indication of key aspects of the proposals. It should be noted that some variation from these recommendations is permitted if it is justified on scientific, veterinary, husbandry, or other welfare grounds. It was the view of the Expert Group that justification on scientific grounds should be specifically authorized by each nation.

Specific Proposals Relating to Dogs

The dog is an inquisitive and highly social animal. The overriding principle is therefore the need to encourage and motivate social housing while providing a complex physical and social environment within the available space. The need for social housing is supported by the association of long-term single housing and social isolation with a range of behavioral disturbances (Hetts and others 1992). The benefits of enriching the environment—both social and physical—have been reported by Hubrecht (1993, 1995) and DeLuca and Kranda (1992). Social interactions are particularly important in dogs from 4 to 20 weeks of age, when social behavior is developing (Scott and Fuller 1965; Scott and others 1974; Wright 1983).

Some key proposals therefore were that:

• Animals should be held in socially harmonious groups with a minimum of two (i.e., a pair).

• Social contacts of puppies between littermates and with humans should be encouraged, particularly during the key socialization period of 4 to 20 weeks of age.

• Separate areas should be provided within pens for different activities (e.g., by the use of raised platforms and pen divisions), should allow for some privacy, and should enable the dogs to exercise some control over their social interactions.

• Physical enrichment items such as dog treats and toys afford welfare benefits, particularly where they meet the dog's chewing behavior and are adequately monitored.

There is a considerable divergence of views on the amount of space necessary for each dog in a pen or other enclosure. The view taken by the Expert Group was that the **minimum space allowances** should take full account of the key points described above and listed in Table 1.

TABLE 1 Minimum Space Allowances Necessary to Encourage Social Housing and Permit Adequate Enrichment of the Environment Regarding Pen Subdivision and Other Factors

Weight of dog (kg)	Minimum floor area for one or two dogs (m²)	For each additional dog add a minimum of (m²)	Minimum height (m)
≤ 20	4	2	2
>20	8	4	2

These recommendations were based on the requirements of **beagles**. For other breeds, space allowances should be determined in consultation with veterinary staff and the national authority. It can be seen from Table 1 that although the smallest pen in which any beagle may be kept should be 4 m², a basic pen unit of 2 m² would allow considerable flexibility.

In deciding space allowances for postweaned stock, the Expert Group took account of the particular needs within commercial breeding establishments and proposed the data that appear in Table 2.

There are differing views on the type of **flooring** appropriate for dogs, and this topic was discussed in detail at the Berlin Workshop (1993). The Expert Group's view was that the preferred flooring for dogs was solid and continuous, with a smooth but nonslip finish, and that open flooring systems such as grids or mesh should be avoided. However, it was decided that there was not sufficient evidence to prohibit open flooring systems, provided they were appropriately designed and constructed, avoided pain, injury, or distress, and allowed the animals to express normal behaviors. Nevertheless, all dogs—on whatever flooring system—should be provided with a comfortable solid resting area within their enclosure. Furthermore, preweaned pups and periparturient and suckling bitches should not be held on an open floor system.

TABLE 2 Space Allowances for Postweaned Dogs

Weight of dog (kg)	Minimum floor area per animal (m²)	Minimum height (m)
≤ 2	0.5	2
>2-3	1.0	2
>5-10	1.5	2
>10-15	2.0	2
>15-20	4.0	2

Specific Proposals Relating to Cats

The proposals on housing and enrichment and on dimensions and flooring for cats took account of consideration similar to those for dogs (Table 3). It was recognized that cats have a strong tendency to learn social behavior, even though they are descended from a solitary species. They do not, for example, form distinct dominance hierarchies. However, it was also recognized that the process of forming social relationships may be stressful and that interpreting visible signs of stress may be more difficult than in dogs.

Key proposals for cats were that:

• They should be socially housed where appropriate; however, social stress in all pair- or group-housed animals should be monitored at least weekly using an established behavioral and/or physiological stress scoring system (e.g., Kessler and Turner 1997).

• Social contacts with littermates and with humans is essential between 2 and 8 weeks of age to encourage the development of social behavior.

• Vertical space should be well utilized and is particularly valuable for cats to provide vantage points, allow for climbing, and allow increased control over their social interactions.

• Pseudopredatory and play behavior should be encouraged, both by providing toys and by interacting with humans. Toys should be changed on a regular basis to avoid familiarity.

As with dogs, **minimum space allowances** for cats took account of the requirement for social housing, for adequate enrichment of the environment by means of subdivisions and provision of enrichment items, and for sufficient separation of areas for different purposes, such as feeding and litter trays.

The question of solid or open **flooring** was considered at the Berlin Workshop (1993). The recommendations agreed by the CoE Working Party were similar for cats and dogs, although the Expert Group decided that cats require a solid floor and should not be kept on open flooring at

TABLE 3 Summarized Proposals of Cage Space Requirements for Cats

	Floor (m^2)	Shelves (m^2)	Height (m)
Minimum for one adult cat	1.5	0.5	2
For each additional cat add . . .	0.75	0.25	Not applicable

any time unless for specific scientific purposes and as authorized by the national authority. The provision of a comfortable resting place (e.g., raised, partly enclosed bed with bedding material) was considered particularly important for this species.

GAPS, SUCCESSES, AND CONCERNS

Guidance on accommodation and care will probably never achieve universal agreement or acceptance, particularly with domesticated/pet species such as dogs and cats, on which a very large number of people would probably claim to be experts but within which there would be a wide range of opinions. There is minimal research-based information available specifically on housing needs, and the Expert Group was able to identify certain areas where focused research would be of benefit (e.g., the relation between minimum space and the quality of the space provided). The possible impact of certain proposals on science also may warrant further study.

Nevertheless, there is a considerable body of data in relation to these species' physiological and ethological needs, which formed the starting point for the Expert Group's discussions. Identifying these needs and then adopting an outcomes-based approach would seem to meet the call for recommendations to be driven by science. Certainly they were driven by the four principles of published data where available, scientific principles, expert opinion, and experience. The inclusion of representation from animal welfare or protection groups provides an example of how the research and animal protection communities can work together to improve animal welfare while recognizing the needs of science. It also enhances political and public acceptability of the process.

The procedure to revise the guidelines that have been adopted by the Council of Europe is a stamina-sapping one, having already occupied more than 5 years. However, this is both a reflection of the extensive and detailed discussion that has taken place and a recognition of the broad range of expert opinions involved. Yet knowledge gained by further research and scientific evidence, as well as changing views on what is currently regarded as good or best practice, will mean that the accommodation and care that should be provided animals in research in the future will make additional revisions of the guidance necessary.

ACKNOWLEDGMENTS

I am very grateful to all members of the Council of Europe's Expert Group on Dogs and Cats for their considerable input related to the proposals. The discussions were held at all times in a positive, constructive, and good-humored way. I am also most grateful to ILAR for inviting me to speak at this international workshop.

REFERENCES

Berlin Workshop. 1993. The accommodation and care of laboratory animals in accordance with animal welfare requirements. Proceedings of an international workshop held at the Bundesgesundheitsamt, Berlin, May 17-19, 1993.

Council of Europe. 1997. European Convention for the Protection of Vertebrate Animals Used for Experimental and Other Scientific Purposes (ETS 123).

De Leeuw., W. 2003. The Council of Europe. In: Proceedings of an ILAR international workshop held in Washington, DC, November 15-17, 2003.

DeLuca, A.M., and Kranda, K.C. 1992. Environmental enrichment in a large animal facility. Lab Anim 21:38-44.

Hetts, S., Clark, J.D., Calpin, J.P., Arnold, C.E., and Mateo, J.M. 1992. Influence of housing conditions on beagle behaviour. Appl Anim Behav Sci 34:137-155.

Home Office. 1989. Code of Practice for the Housing and Care of Animals Used in Scientific Procedures. London: HMSO.

Home Office. 1995. Code of Practice for the Housing and Care of Animals in Designated Breeding and Supply Establishments. London: HMSO.

Hubrecht, R.C. 1993. A comparison of social and environmental enrichment methods for laboratory housed dogs. Appl Anim Behav Sci 37:345-361.

Hubrecht, R.C. 1995. Enrichment in puppyhood and its effect on later behavior of dogs. Lab Anim Sci 45:70-75.

Kessler, M.R., and Turner, D.C. 1997. Stress and adaptation of cats *(Felis silvestris catus)* housed singly, in pairs and in groups in boarding catteries.

NRC [National Research Council]. 1996. Guide for the Care and Use of Laboratory Animals. Washington DC: National Academy Press.

Scott, J.P., and Fuller, J.L. 1965. Genetics and the Social Behaviour of the Dog. Chicago: University of Chicago Press.

Scott, J.P., Stewart, J.M., and De Ghett, V.J. 1974. Critical periods in the organisation of systems. Dev Psychol 7:489-513.

Wright, J.C. 1983. The effects of differential rearing on exploratory behaviour in puppies. Appl Anim Ethol 10:27-34.

Housing for Captive Nonhuman Primates: The Balancing Act

*Melinda Novak**

Nonhuman primates are a crucial research resource because they serve as important models for understanding human health and well-being. However, their success as models depends on balancing two important needs: the research objectives and the well-being of the animals. The balance is not always easy to achieve because we do not fully understand how to maintain primate well-being in a laboratory setting. This discussion reviews information relevant to the housing of rhesus macaques, one of the most commonly used species for laboratory research.

SPECIES-TYPICAL BEHAVIOR

The key to developing effective strategies for housing rhesus monkeys in captivity is to understand their behavior in nature. From a social perspective, rhesus monkeys live in relatively large troops (~ 20-30 individuals on average) that consist of both sexes and all age classes (Lindburg 1971). These troops operate as "closed societies" and repel strangers with acts of aggression (Southwick and others 1965, 1974). Females spend their lives in their natal troop preferentially interacting with their female kin, whereas males usually leave their natal troop during adolescence. Emigrating males spend a period of time either alone or in the company of

---

*This report was presented at the workshop by Dr. Kathryn Bayne, with thanks.

other adolescent males before attempting to gain entry to a new troop. It has been estimated that at least one-third of the males do not survive this change (Berard 1989; Dittus 1979).

Social interactions within the troop are based in part on dominance, wherein some animals have priority of access to incentives. Two features of the dominance hierarchy are redirection of aggression and recruitment of agonistic aid. Threats and aggression can cascade down the hierarchy, with low-ranking animals receiving more "bystander" aggression than higher-ranking animals. Animals threatened or attacked by others frequently attempt to "recruit" others to their defense by screaming at the attacker or by rapidly alternating their gaze between friends and foe (Gouzoules and others 1998).

From an ecological perspective, rhesus monkeys live in a wide variety of different habitats. They have been observed in remote forests, agricultural regions, and many urban areas (Teas and others 1980). Unlike some other primate species, rhesus monkeys appear to thrive in areas of deforestation, and they have been termed "weed" macaques because of this versatility (Richard and others 1989). Rhesus monkeys spend nearly 50% of their time moving to food sites and foraging for food (Goldstein and Richard 1989; Teas and others 1980). They subsist on the fruits and shoots of well over 100 species of plants, and they occasionally supplement their food with eggs, insects, and small animals. This widely varied diet may contribute to their ability to flourish in very different environments.

In addition to these general features, individual monkeys differ with respect to reactive and impulsive temperaments (Suomi 2000). Approximately 20% of the rhesus monkey population appears to be quite reactive to novel events. This reactivity is manifested by heightened and prolonged activation of the hypothalamic-pituitary-adrenal (HPA) axis and by behavioral responses including fear and withdrawal. In contrast, the remaining members of the population show only mild activation of the HPA axis and only brief responses of wariness or caution in response to novel stimuli (Suomi 1991). In nature, a high reactive temperament is associated with heightened emotional responses to maternal disruption (Berman and others 1994), and in males, with later emigration from the natal troop (Suomi and others 1992).

Individual rhesus monkeys also vary with respect to impulsivity. Some male monkeys (~5%) are highly aggressive and do not appear to moderate their aggressiveness with appeasement behavior. This trait is also associated with the presence of low levels of serotonin in the brain as measured by the metabolite 5-hydroxyindoleacetic acid (5-HIAA). In nature, low levels of CSF 5-HIAA in male rhesus monkeys are associated with extreme aggression, earlier emigration from their natal troop compared with other adolescent males (Mehlman and others 1995), and greatly increased risk of mortality (Higley and others 1996).

Knowledge of how monkeys behave in nature can inform how we house and enrich the environments of captive primates. From a social housing perspective, free-ranging monkeys live in complex social groups. Although it is not really possible to duplicate troop life in the laboratory, some form of social housing may be crucial for maintaining well-being. However, there are distinctions that must be considered along with potential costs and benefits. For example, males and females may be affected differently by the presence or absence of partners. Females live in large kin groups throughout their lives, whereas males emigrate and occasionally become solitary. Furthermore, some groupings or pairings will not necessarily be amicable. In nature, rhesus monkey troops are "closed societies," and troop members react aggressively to strangers. Furthermore, social housing may not be optimal or even desirable for certain individuals. In nature, male monkeys with low central nervous system serotonin levels show extreme aggression and are ultimately forced out of their natal troop.

In addition to their complex social environment, rhesus monkeys exist in habitats where they must forage for food and find suitable resting/sleeping sites. Movement and exploration are therefore crucial for survival. Exposure to novel stimuli or foraging devices (i.e., environmental enrichment) would appear to be essential for housing monkeys in captivity; however, this view must be adjusted to account for differences in temperament. Reactive monkeys may show heightened stress responses to enrichment.

LABORATORY FINDINGS

Both social housing and environmental enrichment are considered important regulatory requirements for promoting psychological well-being in captive primates. The logic of this view for rhesus monkeys is derived in part from their life history. However, there are also laboratory studies in which the effectiveness of social housing and environmental enrichment have been examined, and the emerging picture from this work suggests that there are both benefits and costs, depending on the research objectives.

Social Housing

Scientific evidence suggests that there are a number of potential benefits to social housing, the most obvious of which is the ability to groom and affiliate with other monkeys. Companions may also serve as a buffer to stressful events (Winslow and others 2003). Other potential benefits of companionship include increased disease resistance (Shively and others 1989) and improved immune response (Lilly and others 1999; Schapiro and others 2000). However, these relationships are more complex than these descriptions imply. For example, in Shively and colleagues' (1989) study, socially housed female macaques showed less coronary artery disease than singly housed monkeys, but this difference was evident only for dominant socially housed females. Another important benefit of social housing is that there is greater correspondence to the human situation. In a recent study, intracerebroventricular infusions of corticotropin-releasing factor caused depressive-like symptoms, but only in socially housed monkeys (Strome and others 2002).

Social housing is not without cost, and one of the most significant costs is the development of aggression and competition. Rhesus monkeys do not always coexist amicably. Even in stable social groups, aggression can escalate and lead to violent outcomes (Hird and others 1975). From a research perspective, there may be circumstances in which social housing increases experimental variability. For example, moving monkeys from individual cage housing to social housing led to an increase in the availability of dopamine D2 receptors in dominant, but not subordinate, monkeys (Morgan and others 2002). Social housing may also minimize the effects of certain manipulations (e.g., coronary artery disease) or introduce other variables that may mask the effects of manipulations. For example, removal from companions for testing may induce stress reactions as a consequence of separation (see Lyons and others 1998 for squirrel monkeys).

Environmental Enrichment

There are many different methods to enrich the environment of captive primates, ranging from the provision of objects, foraging devices, or videotapes to the redesign of the cage environment (see various commercial cage vendors). As with social housing, the emerging picture suggests that there are both costs and benefits. The most obvious benefit is that environmental enrichment promotes species-typical behavior in the form of exploration. Thus, most monkeys spend some time using foraging devices (Lutz and Novak 1995) and manipulating objects (Novak and others 1993). Some monkeys also appear to watch videotapes (Platt and Novak 1997). The benefits of enrichment may extend beyond mere exploration to include a reduction in stress levels (Boinski and others 1999; Byrne and Suomi 1991) and a decrease in stereotypic behavior (Bayne and others 1991). However, enrichment has not been shown to reduce severe forms of abnormal behavior, such as self-injurious behavior (Novak and others 1998).

Enrichment efforts also incur costs to the animal and to the research enterprise. Rotation of enrichment devices through the colony can increase the risk of disease transmission (Bayne and others 1993). The provision of foraging devices can lead to increased body weight (Brent 1995). Plastic and rubber objects are typically gnawed and chewed, and in some cases can result in injury from foreign material in the intestine (Hahn and others 2000).

Developing optimal housing strategies for rhesus monkeys requires balancing two different but interconnected needs: promoting primate well-being and achieving research objectives. A strong case can be made that both social housing and environmental enrichment foster well-being. However, there are also risks to housing monkeys in social groups and to enriching the environment. Furthermore, the costs and benefits are often relative. What may be a benefit under some conditions can become a cost under other conditions. A thorough review of life history patterns and a careful cost-benefit analysis may provide guidance in designing housing strategies for particular research programs.

REFERENCES

Bayne, K.A., Dexter, S.L., Hurst, J.K., Strange, G.M., Hill, E.E. 1993. Kong toys for laboratory primates: Are they really an enrichment or just fomites? Lab Anim Sci 43:78-85.

Bayne, K., Mainzer, H., Dexter, S., Campbell, G., Yamada, F., Suomi, S.J. 1991. The reduction of abnormal behaviors in individually housed rhesus monkeys *(Macaca mulatta)* with a foraging/grooming board. Am J Primatol 23:23-35.

Berard, J.D. 1989. Life histories of male Cayo Santiago macaques. Puerto Rico Health Sci J 8:61-64.

Berman, C.M., Rasmussen, K.L.R., Suomi, S.J. 1994. Responses of free-ranging rhesus monkeys to a natural form of social separation: I. Parallels with mother-infant separation in captivity. Child Dev 65:1028-1041.

Boinski, S., Swing, S.P., Gross, T.S., Davis, J.K. 1999. Environmental enrichment of brown capuchins *(Cebus apella):* Behavioral and plasma and fecal cortisol measures of effectiveness. Am J Primatol 48:49-68.

Brent, L. 1995. Feeding enrichment and body weight in captive chimpanzees. J Med Primatol 24:12-16.

Byrne, G.D., and Suomi, S.J. 1991. Effects of woodchips and buried food on behavior patterns and psychological well-being of captive rhesus monkeys. Am J Primatol 23:141-151.

Dittus, W.P.J. 1979. The evolution of behavior regulating density and age-specific sex ratios in a primate population. Behaviour 69:265-302.

Goldstein, S.J., and Richard, A.F. 1989. Ecology of rhesus macaques *(Macaca mulatta)* in Northwest Pakistan. Int J Primatol 10:531-567.

Gouzoules, H., Gouzoules, S., Tomaszycki, M. 1998. Agonistic screams and the classification of dominance relationships: Are monkeys fuzzy logicians? Anim Behav 55:51-60.

Hahn, N.E., Lau, D., Eckert, K., Markowitz, H. 2000. Environmental enrichment-related injury in a macaque *(Macaca fascicularis)*: Intestinal linear foreign body. Comp Med 50:556-558.

Higley, J.D., Mehlman, P.T., Higley, S.B., Ferrnald, B., Vickers, J., Lindell, S.G., Taub, D.M., Suomi, S.J., Linnoila, M. 1996. Excessive mortality in young free-ranging male non-human primates with low cerebrospinal fluid 5-hydroxyindoleacetic acid concentrations. Arch Gen Psychiatry 53:537-543.

Hird, D.W., Henrickson, R.V., Hendrickx, A.G. 1975. Infant mortality in *Macaca mulatta*: Neonatal and postnatal mortality at the California Primate Research Center, 1968-1972. A retrospective study. J Med Primatol 4:4-22.

Lilly, A.A., Mehlman, P.T., Higley, J.D. 1999. Trait-like immunological and hematological measures in female rhesus across varied environmental conditions. Am J Primatol 48:197-223.

Lindburg, D.G. 1971. The rhesus monkey in North India: An ecological and behavioral study. In: Rosenblum, L.A., ed. Primate Behavior Developments in Field and Laboratory Research. New York: Academic Press. p. 1-106.

Lutz, C.K., and Novak, M.A. 1995. The use of foraging racks and shavings as enrichment tools for social groups of rhesus monkeys *(Macaca mulatta)*. Zoo Biol 14:463-474.

Lyons, D.M., Kim, S., Schatzberg, A.F., Levine, S. 1998. Postnatal foraging demands alter adrenocortical activity and psychosocial development. Dev Psychobiol 32:285-291.

Mehlman, P.T., Higley, J.D., Faucher, I., Lilly, A.A., Taub, D.M., Vickers, J., Suomi, S.J., Linnoila, M. 1995. Correlation of CSF 5-HIAA concentration with sociality and the timing of emigration in free-ranging primates. Am J Psychiatry 152:907-913.

Morgan, D., Grant, K.A., Gage, H.D., Mach, R.H., Kaplan, J.R., Prioleau, O., Nader, S.H., Buchheimer, N., Ehrenkaufer, R.L., Nader, M.A. 2002. Social dominance in monkeys: Dopamine D2 receptors and cocaine self-administration. Nature Neurosci 5:169-174.

Novak, M.A., Kinsey, J.H., Jorgensen, M.J., Hazen, T.J. 1998. The effects of puzzle feeders on pathological behavior in individually housed rhesus monkeys. Am J Primatol 46:213-227.

Novak, M.A., Musante, A., Munroe, H., O'Neill, P.L., Price, C., Suomi, S.J. 1993. Old socially housed rhesus monkeys show sustained interest in objects. Zoo Biol 12:285-298.

Platt, D.M., and Novak, M.A. 1997. Video-stimulation as enrichment for captive rhesus monkeys *(Macaca mulatta)*. J App Anim Behav Sci 52:139-155.

Richard, A.F., Goldstein, S.J., Dewar, R.E. 1989. Weed macaques: The evolutionary implications of macaque feeding ecology. Int J Primatol 10:569-594.

Schapiro, S.J., Nehete, P.N., Perlman, J.E., Sastry, K.J. 2000. A comparison of cell-mediated immune responses in rhesus macaques housed singly, in pairs, or in groups. Appl Anim Behav Sci 68:67-84.

Shively, C.A., Clarkson, T.B., Kaplan, J.R. 1989. Social deprivation and coronary artery atherosclerosis in female cynomolgus monkeys. Atherosclerosis 77:69-76.

Southwick, C.H., Beg, M.A., Siddiqi, M.R. 1965. Rhesus monkeys in North India. In: DeVore, I., ed. Primate Behavior Field Studies of Monkeys and Apes. New York: Holt, Rinehart and Winston. p. 111-159.

Southwick, C.H., Siddiqi, M.F., Farooqui, M.Y., Pal, B.C. 1974. Xenophobia among free-ranging rhesus groups in India. In: Holloway, R., ed. Primate Aggression, Territoriality, and Xenophobia: A Comparative Perspective. New York: Academic Press. p. 185-209.

Strome, E.M., Wheler, G.H., Higley, J.D., Loriaux, D.L., Suomi, S.J., Doudet, D.J. 2002. Intracerebroventricular corticotropin-releasing factor increases limbic glucose metabolism and has social context-dependent behavioral effects in nonhuman primates. Proc Natl Acad Sci U S A 99:15749-15754.

Suomi, S.J., 1991. Up-tight and laid-back monkeys: Individual differences in the response to social challenges. In: S. Brauth, W. Hall, and R. Dooling, eds. Plasticity of Development. Cambridge: MIT Press. p. 27-56.

Suomi, S.J., 2000. Behavioral inhibition and impulsive aggressiveness: Insights from studies with rhesus monkeys. In: L. Balter, C. Tamis-Lamode, eds. Child Psychology: A Handbook of Contemporary Issues. New York: Taylor and Francis. p. 510-525.

Suomi, S.J., Rasmussen, K.L.R., Higley, J.D. 1992. Primate models of behavioral and physiological change in adolescence. In: E.R. McAnamey, R.E. Kriepe, D.P. Orr, and G.D. Comerci, eds. Textbook of Adolescent Medicine. Philadelphia: Saunders. p. 135-139.

Teas, J., Richie, T., Taylor, H., Southwick, C. 1980. Population patterns and behavioral ecology of rhesus monkeys *(Macaca mulatta)* in Nepal. In: Lindburg, D., ed. The Macaques Studies in Ecology, Behavior and Evolution. New York: Van Nostrand Reinhold Company. p. 247-262.

Winslow, J.T., Noble, P.L., Lyons, C.K., Sterk, S.M., Insel, T.R. 2003. Rearing effects on cerebrospinal fluid oxytocin concentration and social buffering in rhesus monkeys. Neuropsychopharmacology 28:910-918.

Assessment of Animal Housing Standards for Rabbits in a Research Setting

Markus Stauffacher and Vera Baumans

INTRODUCTION

Laboratory animals such as rabbits are bred and housed for experimental use. The living conditions, housing, and husbandry are often more obstructive and more stressful for the animals than the experimental procedure itself. Therefore, the potential negative effects of an experiment on a laboratory animal's well-being are not restricted to the experiment itself but instead cover the whole life span of the animal.

Discussions of welfare requirements and their practical implementation could be improved substantially if decision makers would bear in mind that the desire to protect animals in captivity is based on ethical considerations of humans. However, in contrast to this perspective, the true well-being of captive animals should be based on a biological understanding that relates to the specific needs of the respective species and strains.

In 1986, the current housing standards for laboratory rabbits were established in Article 5 of the Council of Europe's *Convention for the Protection of Vertebrate Animals Used for Experimental and Other Scientific Purposes* (CoE 1986): "Any animal used or intended for use in a procedure shall be provided with accommodation, an environment, at least a minimum degree of freedom of movement, food, water and care, appropriate to its health and well-being. Any restriction on the extent to which an animal can satisfy its physiological and ethological needs shall be limited as far as practicable. In the implementation of this provision, regard should be paid to the guidelines for accommodation and care of animals set out in Appendix A to this Convention. . . ." Cages and pens "should be designed for the well-being of the species" and "should permit the satisfaction of certain ethological needs (for example the need to climb, hide or shelter temporarily). . ." (CoE 1986, Appendix A, paragraph 3.6.3).

However, on the species-specific level, the guidelines are restricted to recommendations on minimum cage dimensions and stocking densities. Regulations set up by a political body do not define an optimum but instead set limits and minimum standards. All concepts of animal protection are composed of conventions and assessments that are inevitably linked to those individuals who prepare and make the decisions. Working out minimum requirements with respect to animal welfare (ethical) and to the supposed well-being of laboratory animals (biological) is, last but not least, a political (mostly economical) question. Nevertheless, the decision-making process must be based first and overall on sound arguments concerning the biology of species and strains in question.

The environment of a rabbit kept in captivity—as a pet, for fattening, or in the laboratory—has a considerable impact on its well-being and functioning. Important environmental factors include not only climate (e.g., light cycle, temperature, relative humidity, ammonia concentration, and ventilation) but also hygiene, food and water supply, housing, and the presence of conspecifics. In this presentation, we focus on the housing environment.

ESTABLISHED SPACE REQUIREMENTS

The minimum space requirements of the European Convention (CoE 1986) are based on a mathematical calculation model with arbitrarily set constant factors, slope, and starting point. The slopes represent weight-bands and allow the space requirements for a given number of rabbits to be calculated. The heavier the rabbits, the fewer square centimers of space are required per weight unit. The calculation model refers only to body weight and does not make a distinction between strains, sex, and age. As a result, this model does not adequately reflect the fact that young, growing animals need much more space in relation to their body weight than adults.

The minimum space requirements of the European Convention ETS 123, Appendix A, 1986, apply to a medium-sized (< 4 kg) rabbit such as the New Zealand White rabbit: 2500 cm^2 with a height of 35 cm. However, in the Sixth Edition of the *Guide for the Care and Use of Laboratory Animals* (the *Guide*) (NRC 1985), the requirement applies to a rabbit of 2700 cm^2 with a height of 35 cm.

When we consider the current standard housing of laboratory rabbits from a human point of view, single-housed animals are easy to control and handle. Animal staff can easily handle a cage size of approximately 2500 cm^2 with a height of 35 cm and can easily disinfect the metal and plastic walls. The slatted or perforated floors allow automatic cleaning, and the pelleted food and water bottles can be controlled. All of these factors are consistent with Good Laboratory Practice.

In contrast, from the rabbit's perspective, the current housing practices provide the following characteristics: limited freedom of movement, a barren cage environment with restricted challenges and possibilities for occupation, no social partners, and an artificial open nest-box. All of these characteristics afford the rabbit hardly any possibility of performing species-specific behavior.

A series of studies have shown that in cages that comply with the minimum dimensions required in Appendix A of the European Convention (CoE 1986), the welfare of laboratory rabbits is impaired (Stauffacher 2000). The consequences of limited freedom of movement are changes in locomotor patterns and sequences (e.g., inability to hop), which result in skeletal damage in, for example, the femur proximalis and the vertebral column (Bigler 1995; Drescher 1993). The barren cage environment with a severe lack of stimulation leads to behavioral disorders such as wire-gnawing and excessive wall-pawing, as well as to panic reactions and signs of "boredom" (Lidfors 1997). During breeding, an open nest-box and poor quality and quantity of nesting material do not permit the natural behavior of the doe (e.g., closing up the nest entrance when triggered by odor cues of the litter). In addition, these conditions do not allow the doe any chance to withdraw from the litter, which can result in behavioral disorders in the mother and in significant rearing losses (Stauffacher 2000).

In 1997, the Multilateral Consultation of the Council of Europe adopted a resolution on the accommodation and care of laboratory animals, which specified that "young and female rabbits should be housed in socially harmonious groups . . ." and that "pens, as well as cages, should include enrichment material e.g. roughage, sticks, an area for withdrawal and nesting material." As a consequence of that resolution, an international expert group was set up in 1998 to devise science-based proposals for a revision of Appendix A.

RABBIT BEHAVIOR AND NEED FOR SPACE

Rabbits do not use space per se; they use resources and structures within an area for specific behaviors. Appropriate structuring of the cage/pen environment may be more beneficial than provision of a larger floor area; however, a minimum floor area is needed to provide a structured space (e.g., blinds, shelters, and platforms) that includes withdrawal areas and vantage points.

Rabbits tend to be highly motivated, to make use of enrichment based on food items, and to satisfy their need for roughage (hay, straw) and gnawing (soft wood chew sticks). With respect to the social environment of rabbits, the domestic rabbit with its wild ancestor the European wild rabbit (*Oryctolagus cuniculi* L.) is a highly social animal that lives in the wild in stable groups of one male, three to five females, and their offspring in a home territory or "warren." They establish a linear rank order (males and females), and subadult males must leave the warren.

A social partner always creates new and unpredictable situations to which a rabbit must react. Such situations lead to an increase of alertness and exploratory behavior, and provide diversion, occupation, and probably also some feelings of "security." References to these situations appear in the following documents: (1) The Council of Europe Multilateral Consultation on the Revision of the Convention for the Protection of Animals Used for Scientific or Other Purposes, ETS 123, Part II, Appendix A, 1997: "Young and female rabbits should be housed in socially harmonious groups unless the experimental procedure or veterinary reasons make this impossible." (2) The Seventh Edition of the *Guide* (NRC 1996): "Animals should be housed with a goal of maximizing species-specific behaviors and minimizing stress-induced behaviors. For social species, this normally requires housing in compatible pairs or groups."

The crucial question, however, is how to assess minimum recommendations. To determine the minimum recommendations for sizes of primary enclosures (cages, pens) for laboratory rabbits, it is necessary to consider both the quantity and the quality of space. The crucial point is the interaction between the space—the structure of the cage, the rabbits, and the type and quantity of enrichment provided. These variables must be based on experimental results and scientific papers, good/best practice, and experimental constraints. Although good scientific arguments may exist regarding why limits should be set in particular cases, the exact numeric values for minimum cage sizes and heights as well as for maximum stocking densities can never be scientifically evaluated and "proved."

RECOMMENDATIONS AND ANTICIPATED REVISIONS

In the proposal for the revision of Appendix A of the European Convention ETS 123, the Expert Group on Rodents and Rabbits recommends that medium-sized (< 4 kg) rabbits such as New Zealand White rabbits should be housed in cages with a floor area of 4200 cm^2 and a height of

45 cm, including a raised area of approximately 55 × 30 cm where one rabbit or two compatible rabbits can be housed. Without a raised area, floor space should be 5600 cm² for one rabbit and 6700 cm² for two rabbits. According to Swiss law, one rabbit or two compatible rabbits can be housed in a cage with a floor that measures 4200 cm² + 1800 cm² with a height of 60 cm (SOAP 1991).

In the process of refinement of housing standards, the following points should be taken into account:

1. Biological facts and scientific evidence (related to the animals);
2. Experimental tasks and constraints (related to the research goals);
3. Practical experience (related to the debating subjects);
4. Ethical principles (related to animal protection); and
5. Assessment of economical and political reasonableness (related to human societies).

REFERENCES

Bigler, L. 1995. Zusammenfassung der Ergebnisse radiologischer Untersuchungen an Zuchtzibben-Wirbelsäulen und Mastkaninchen-Femurknochen in der Schweiz von 1984-1995 [in German]. Bern: Bericht z.Hd. Bundesamt für Veterinärwesen.

CoE [Council of Europe]. 1986. European Convention for the Protection of Vertebrate Animals Used for Experimental and Other Scientific Purposes (ETS 123). Strasbourg: Council of Europe.

Drescher, B. 1993. Zusammenfassende Betrachtung über den Einfluss unterschiedlicher Haltungsverfahren auf die Fitness von Versuchs- und Fleischkaninchen [in German]. Tierärztl Umschau 48:72-78.

Lidfors L. 1997. Behavioural effects of environmental enrichment for individually caged rabbits. Appl Anim Behav Sci 52:157-169.

NRC [National Research Council]. 1985. Guide for the Care and Use of Laboratory Animals. 6th Ed. Washington, DC: National Academy Press.

NRC [National Research Council]. 1996. Guide for the Care and Use of Laboratory Animals. 7th Ed. Washington, DC: National Academy Press.

SOAP [Swiss Ordinance on Animal Protection]. 1991. Schweiz. Tierschutzverprdnung vom 27. Mai 1081, Änderung vom 23. Oktober 1991. Bern: Eidgenössische Daten-und Materialzentrale, 1991.

Stauffacher, M. 2000. Refinement in rabbit housing and husbandry. In: M. Balls, A.-M. Van Zeller, M.E. Halder, eds. Progress in the Reduction, Refinement and Replacement of Animal Experimentation. Amsterdam: Elsevier Science B.V. Publ. p 1269-1277.

Session 3

Approaches to Current Guidelines— United States and Europe

Housing Standards: Development of Guidelines and the Process for Change

William J. White

The 1996 *Guide for the Care and Use of Laboratory Animals* (*Guide*) is the most recent version of seven editions of the document beginning with the 1963 edition. The 1963 *Guide* was developed by a committee of seven members and consisted of 33 pages divided into three sections, whereas the 1996 *Guide* was developed by a committee of 16 members and spanned 125 pages divided into five sections. The charge to the 1996 *Guide* committee was to develop a guidance document for laboratory animal care and use—not to develop regulations.

COMMITTEE FOR REVISION OF THE *GUIDE*

The writing of the 1996 *Guide* spanned five committee meetings and involved seven major drafts with numerous minor drafts developed over a 2-year period. The major sections of the *Guide* were selected based on the principal components of an animal care and use program. Subcommittees of the parent committee prepared drafts of sections for full committee review and discussion. Literature searches were provided by the National Agricultural Library. The final document underwent two rounds of external review before being published.

The charge to the committee and the committee's approach set the tone for the document. It was thought that previous guidelines were accepted and generally were serving well; hence the committee was charged with updating and improving, as well as addressing any shortcomings. It was believed that radical departure would probably be difficult to justify, and the committee wanted to be certain that the guidance was compatible with existing regulations in the United States. It was clear to the committee that a broad vision was necessary because the document would have to be applied to diverse units, settings, and situations. The recommendations made in the *Guide* had to be balanced between science, animal well-being, ethics, and resource requirements. The committee believed it was essential to reaffirm the institutional animal care and use committee (IACUC) as the principal local oversight mechanism and empower it to administer a performance-based approach.

COMMUNICATING THE PERFORMANCE-BASED APPROACH

The performance-based approach specifies the desired outcome and provides criteria for assessing the outcome, but it does not specify how to achieve the outcome. In constructing the *Guide,* it was appropriate in some instances to provide examples, but to emphasize clearly that multiple methods could be acceptable and the choice of method had to be adjusted, based on circumstances. Given the lack of data on which to base detailed guidance in many aspects of animal care and use, the performance-based approach used in the *Guide* encouraged the development of data that could be used to improve animal care and use.

The committee began the process by a careful review of the existing guidance and regulations. It considered whether there were problems with existing guidance and whether new science or technology was available that needed to be addressed. It examined the literature, especially that generated since the last *Guide* was published, to determine whether new peer-reviewed literature provided significant information that required inclusion. The committee thought it was important to use all sources of information in constructing the *Guide.* In particular, opinion and data from the public, as well as the affected community, were actively sought through public meetings and written comments. Every committee member attended the public meetings and read all written materials submitted.

The quality of information available to the committee varied, from statements of opinion, to unpublished reports, to comprehensive peer-reviewed scientific studies. The committee analyzed the peer-reviewed literature, which varied considerably in adequacy and approach. In some cases, only a single study under a defined set of conditions using limited measures was available. In other cases, a series of studies with multiple measures exploring a range of conditions or practices could be found, whereas in other cases, published reports could be found that surveyed large numbers of animals with limited measures and limited control.

Guidance developed by the committee had to be qualified, based on the quality of the information available. Of particular concern was the general applicability of findings across multiple strains, age, sex, and species. In addition, the magnitude of the changes found and their potential impact on study variation and animal well-being had to be weighed carefully. Some studies used insensitive measures that could be confounded by other variables. Data from such studies were interpreted to have some value, but recommendations had to be tempered. Often there were only a limited number of studies that directly addressed a particular topic, and these studies often were unlinked or in some cases were conflicting. In a number of cases, studies could be classified as proof of principle, but with no exploration of mechanism or their general applicability, and they seldom could be verified independently. In a number of instances, professional judgment had to fill in gaps where studies were lacking. Overall, however, if there was little evidence available, the committee avoided the temptation to make assumptions and to extend conclusions beyond available data. Of concern to the committee was that any guidance had to fit practically and logically into existing animal care programs and had to be of sufficient impact/importance to suggest action.

The committee spent time analyzing how new guidance would apply to different animal care and use situations. It was important to assure that the guidance made sense and, if there were exceptions, to indicate how common they were and why they occurred. The committee also tried to determine whether the proposed changes in guidance would generate unintended consequences. Thus, if a guidance recommendation improved one aspect of animal well-being at the expense of another, such as causing a potential increase in animal usage by the application of some forms of environmental enrichment, it was necessary for the committee to determine whether the trade-off was appropriate.

IMPACT OF RECOMMENDATIONS

The committee was also concerned about the impact of recommendations on the conduct of research or testing including a potentially increased risk of microbiological contamination. Implementation of recommendations that might contribute to microbiological contamination could invalidate many types of research and thus result in greater animal usage. The guidance provided in the *Guide* had to be achievable, such that the intended benefit would be proportional to the resources required. The committee felt that any changes proposed in the *Guide* had to be defensible at least in proportion to their specificity. For example, if cage space allocated per rat of a given weight was currently 23 in^2 and it was suggested to change it to 30 in^2, why was 28 or 35 in^2 not chosen? It was clear that more than good intent was required to make changes—in fact, good data were needed to be specific in making such changes.

The committee believed that the guidance had to be very clear. Any recommendations had to use terms that were clearly defined either within the document or within appropriate references. In many cases, the committee provided examples to clarify intent and to expand the meaning of terms. The committee also thought it appropriate to ensure that the characteristics of intended outcomes of the recommendations were clearly described. Overall, the tone of the document was an explanatory one in which the rationale behind the recommendations was given in some detail. The committee believed that wording was critical in developing the document. It avoided poorly defined or emotional terms. It used words to indicate importance as well as limits of application or knowledge. It reserved the word *must* for programmatic issues for which there was no other interpretation or method. *Must* was also used where there was overwhelming scientific information or ethical considerations. There are very few *musts* in the 1996 *Guide.*

The word *should* was used as a strong recommendation for achieving a goal, but it was clearly recognized that some individual circumstances justified an alternative strategy. Words such as *can, might, could, recommend,* and *encourage* were considered alternative verbs to indicate that the recommendations may have to be modified, multiple methods of achieving the outcome were possible, only limited information was available, or that the guidance may only apply to certain circumstances.

PRACTICAL ISSUES IN THE DEVELOPMENT OF GUIDANCE FOR HOUSING STANDARDS

One of the most difficult sections to construct in the *Guide* was the animal environment section, which included housing needs of laboratory animals. Although it was intuitive that the housing environment of laboratory animals somehow affects their performance and well-being, there was not a great deal of information available to provide very specific recommendations. In most cases, only unrelated proof of principle studies that demonstrated housing/environmental effects on animals or research results were available. In a number of studies, there were clear design flaws, which included the use of small numbers of animals, the use of a single species, stock, or strain, and the failure to dissect out confounding variables such as the effects of group size and animal density with respect to cage space effects.

Temperature and Humidity

A review of housing guidelines and published data pertaining to temperature and humidity reveals the gaps in our knowledge that make it difficult to specify precise environmental conditions. Most common laboratory animals are adaptive homeotherms and as such make anatomic, metabolic, and physiologic adjustments in response to their environment to maintain well-being. Environmental adaptation in both wild and laboratory animals suggests that consistency in environment may not be "normal" or perhaps even desirable. To demonstrate effects caused by temperature or relative humidity, it may be necessary to have a complex/unique set of conditions present unless extreme and clearly unacceptable conditions are utilized. For example, testicular degeneration/infertility in mice has been shown to occur when temperatures exceed 83°F within the secondary enclosure.

It would appear that the rationale for specifying any temperature or relative humidity conditions within the laboratory, other than avoiding extreme conditions that clearly could cause harm to the animals, would be to control research variation caused by unpredictable adaptation to housing conditions. The adaptive processes may, in fact, serve the animals quite well, but those are the processes that may interfere with research results. The question then remains as to how much variation due to adaptive processes is acceptable and, hence, what limitations must be placed on temperature and relative humidity. It also begs the question of why such changes would not be sorted out by the use of appropriate controls. Clearly, a number of variables would affect these adaptive processes, including the type of housing (e.g., pen, run, open cage, microisolation cage, isolator), the type of ventilation system used within the primary and secondary enclosures, the specifics of cage/room coupling of ventilation, as well as stratification of temperature and relative humidity within the room itself.

Interaction of other environmental factors with the thermal regulatory behavior of rodents is also an important consideration. Substantial existing data demonstrate that singly housed mice prefer ambient temperatures between 28°C and 30°C whereas group-housed animals prefer temperatures between 24°C and 27°C. The ability of group-housed animals to share metabolically generated heat by huddling together explains the differences in selected ambient temperatures between group- and singly housed animals. The resultant effective ambient temperatures in both group- and singly housed mice are compatible with the estimates of thermal neutral zones for mice of 28°C to 30°C. This range is at variance with the human comfort zone of 22°C (± 2°C), which appears to correspond more closely to temperatures recommended within guidance documents.

Bedding

Recent studies have shown that bedding types that allow burrowing or nesting allow operating ambient temperatures to be increased from an ambient of 22°C to an effective ambient of 29°C. By contrast, bedding that allowed only resting on its surface, but not burrowing, increased temperatures by 2 to 4°C depending on the bedding type. These findings beg the question of whether rodents are using bedding for thermal regulation rather than for psychological enrichment, as suggested in some guidance documents. It is possible that we are recommending the correct thing in terms of the use of bedding or nesting materials for the wrong reasons.

Cage Space

Cage space is another environmental parameter for which there are very few specific data. Providing an adequate amount of cage space is generally thought to be important for animal well-being. Unfortunately, guidelines have been established based on consensus, surveys, some data, and appearance to observers. Although it is intuitive that relationships must exist between cage space and parameters thought to indicate over-crowding, these relationships have never been studied well and are prone to be influenced by a wide range of variables. Most studies have not separated out the effects of group size (number of animals in an enclosure) from density (space provided to each animal regardless of the number in the group). These effects are independent of each other and may be affected by sex and the age of the animals in the enclosure. In general, single housing appears to be the most likely to elicit negative effects on the animals. Recommendations for cage space were developed in the early 1960s and published in the first ILAR *Guide* as suggestions, not standards. In 1969, a density-based set of recommendations using six weight categories was developed in all likelihood from a survey of common practices coupled with limited unpublished data. Although this guidance was modified slightly in ensuing years, principally by adjusting weight ranges and providing for very large animals, the body weight/space relationship remains almost linear, which is suspicious. European regulations currently in effect express this same relationship simply, and often use a continuous rather than a discontinuous format. Newly proposed European regulations deviate significantly from this space allocation, but the changes do not appear to be based on peer-reviewed literature specific to species and weight.

Significant data exist to establish that rodents are thigmotaxic, i.e., they prefer to be along the edges of cages rather than in the exposed center of the cage. Moreover, they appear to utilize hiding places/shelters.

Specifying cage space requirements by virtue of density guidelines appears to be contrary to these natural behaviors because it disproportionately increases the unused center portion of the cage versus the used perimeters of the cage. It is quite probable that cage space, like temperature, cannot be specified very tightly and that a range of available floor space or usable surface is acceptable depending on the type and quality of the space. There is clearly a need to explore requirements of rodents and other animals with respect to group housing because it appears that there are complex relationships that differ with group size. Sufficient evidence exists to suggest that enclosure design or complexity can alter space requirements.

NEED FOR ADDITIONAL SCIENTIFIC DATA

Clearly, much more information needs to be generated before any changes are made in guidance. Complex interactions need to be more clearly understood and considered across a wide range of applications of these guidance documents. It is unlikely that a very defined amount of critical space can be shown to be an absolute requirement, and it is unlikely that relationships are going to be linear. Key information is not available on topics as simple as occupied floor area versus body weight. It is difficult to conceive how guidance documents can be updated without a great deal more peer-reviewed information. At the very least, in generating such information, more than proof of principle studies alone need to be done, and it is imperative that multiple parameters be measured and that both positive and negative controls be provided. It is also essential that there be consideration of confounding variables and confirmation of findings under field conditions. There should be some ranking of physiologic/metabolic and behavioral significance because simply describing that a condition exists and is statistically significant may not be adequate justification for providing for all eventualities in animal housing.

Revision of Appendix A to the European Convention ETS 123: The Participants, the Process, and the Outcome

Derek Forbes

Papers presented by earlier speakers have emphasized that the "Guidelines for accommodation and care of animals" presented in Appendix A of the Convention have proven to be very useful and have been widely applied. However, since 1976 when the Convention was first applied, scientific knowledge and experience have advanced considerably. Moreover, there has been an increased public interest in and awareness of animal usage in experimentation and a regard for their welfare. The author as a representative of a nongovernmental organization (NGO; see below) recognizes and acknowledges the importance of the participation and input from the wide spectrum of interested parties who have contributed to the process in satisfying, to the extent possible, the needs of animals used in research, those who work with them, and the public on whose behalf the work is done.

The participants who are involved in the Working Party preparing for Multilateral Consultation of Parties to the Convention include the following:

• Parties: Belgium, Netherlands, Cyprus, Norway, Czech Republic, Spain, Denmark, Sweden, Finland, Switzerland, France, United Kingdom, Germany, European Community, and Greece.
• Signatory States: Bulgaria, Ireland, Portugal, Slovenia, and Turkey.
• Observers—Member States: Austria, Croatia, Hungary, Italy, and Malta.

- Observers—Nonmember States: Australia, Canada, Holy See, Japan, New Zealand, and the United States of America.
- Participants Who Are Experts from International Organizations: Canadian Council on Animal Care, European Biomedical Research Association, European Federation of Animal Technologists, European Federation for Primatology, European Federation of Pharmaceutical Industries and Associations, European Science Foundation, Eurogroup, Federation of European Laboratory Animal Breeders Associations, Federation of European Laboratory Animal Science Associations, Federation of Veterinarians of Europe, Institute for Laboratory Animal Research, International Council for Laboratory Animal Science, International Society for Applied Ethology, and World Society for Protection of Animals.

An examination of the list reveals first that all the organizations represented were nongovernmental, pan-European, or internationally recognized. Second, the organizations included those whose primary concerns were research based, or who had expert knowledge of the science and care of animals used in research, as well as those who concentrated on the protection of animals used in the laboratory and the ethological restrictions that such use imposed on the animals. Although the interests may appear diverse, experience gained from working alongside persons from such groups has shown that everyone involved was very aware of the paramount need to satisfy the animal's welfare and ensure its well-being.

During the course of the revision, the process has developed and evolved in content from that which was included in the original Appendix A. The strategy was determined within the working party by the member states. Initially, the general part of the appendix was updated. Thereafter the sections dealing with species of animals most commonly used in research and that had been included in the original convention were revised. However, it was realized that other animals were used in research for which there were no agreed-upon standards within Europe. This realization led to the decision to include all of the following groups within the remit of the revision. The full list of species now includes rodents and rabbits; dogs, cats, and ferrets; nonhuman primates; birds; farm animals (sheep, goats, cattle, horses, (mini)pigs); fishes; and amphibians and reptiles.

A group of experts drawn from the NGOs considered each of the groups listed. The constitution of each group included individuals representing the diverse interests of the experts as described above. The format inevitably was conducive to change because it was necessary to reach some consensus within each of the working groups. The overriding prerogative was to try to achieve an enriched environment that satisfied the ethological needs of the animal, with special attention being given to the most appropriate implementation for the species concerned. Because most of the species are social animals, it was agreed that interaction with conspecifics by group housing was of paramount importance. Although some form of containment in accommodation is usually inevitable, the best possible utilization of space in relation to the animal's natural activities was a second important criterion. Some variety within the environment was also crucial to provide stimulation.

Each group of experts produced a report on one or more species that was then presented to the next meeting of the Working Party. There it was debated and amended, or additional information was requested, before it was submitted back to the expert group for their further consideration. In some cases, this cycle has been repeated several times, often with each stage reversing or revising issues discussed at the previous meeting. It should be noted that during the protracted process, there has been a war of attrition between interest groups, the outcome of which has been an eventual meeting of minds in consensus and the adoption of standards that are reasonable and accepted.

Faster progress was made after the introduction of a Drafting Group. This group was composed of a small number of representatives of the national authorities together with the Secretariat. The group met between meetings of the Working Party and rationalized the output of the previous meeting. The groups of experts produced reports with varying formats that the Drafting Group standardized, which greatly facilitated progress. For example, all of the environmental standards that were common to all species were described in the general part of the Appendix, leaving only those specific to a particular group to be mentioned in the species-specific text.

During the course of its several meetings, the Working Party has progressively "finalized" most of the species-specific reports as well as the General Part. Such finalization denotes that the document concerned will not be open for any further discussion. This discipline has been essential because the discussions could have been endlessly iterative, considering the breadths of opinions represented. Everyone involved in the process has been aware of the paucity of good scientifically based data that could be used to optimize the environment of animals used in research. They also recognize the difficulty of producing such data. Therefore, the outcome of the process will be a consensus of those with a genuine interest and knowledge in the subject. Although it will result in changes that will have a financial cost, the process overall should be seen as the best way to satisfy the needs of science and politics in improving the welfare of animals used in essential research.

Breakout Session: Rats and Mice

Leader: Axel Kornerup-Hansen

Rapporteur: Rosemary Elliott

The purpose of this session was to elicit different points of view, based on the following set of questions and issues introduced by the breakout leader. Brief comments of participants during the general discussion are provided below. Two of the participants in this session, Abigail Smith and Jan Ottesen, were invited by the breakout leader to present the results of studies conducted in their laboratories. The summaries of these presentations also appear in this section.

Why change the guidelines?

What is the basis for creating new guidelines?
- Scientific proof
- Experience
- Good/best practice

Define good/best practice.
- The European guidelines deal with these issues.
- Space for mice changes only for young ones, which have increased space.
- Space for rats shows more change, increasing 35%, especially during experiments, primarily as an increase in average height. The rationale for this observation was not that more space creates better welfare, but allows for enrichment.

Is there a need to change existing guidelines for rats and mice?

The most important issues to be considered are enrichment, solid flooring, and social housing.

COMMENTS FROM AUDIENCE

Several members of the audience were rather skeptical about several aspects of enrichment. It was suggested that many enrichment items, such as houses and tubes, can be used without increasing the space. Because these objects also can provide extra space for exploration of the mice or rats, they themselves increased the useable space. The assumption that enrichment increases an animal's well-being was questioned. It was suggested by participants that scientific data should be sought in order to avoid assumptions. Along these lines it was suggested that animals be observed for a 24-hour period to determine the use made of the enriching items and to attempt to assess their value. Another suggestion was to rearrange the space, by adding structures. One cage designer indicated that he keeps floors clear, but adds a resting shelf or feeding structure.

One speaker expressed concern about how the meeting was progressing and would like to see more science-based knowledge on all issues. Another speaker introduced the issue of the impact on the science in which the animals are used, and asked whether scientists had had input on the decisions. There was discussion on defining an optimal enrichment device and whether commercial breeders were expected to use enrichment and whether additions to bedding such as nestlets were acceptable.

Several speakers expressed concern about what happens to animals during data collection, in contrast to animals in breeding and stock colonies. For instance, what would be the effect of the lack of enrichment objects in metabolic cages, particularly for animals used to enrichment? Others wished to know the effect of modifying cages on experimental results and whether one should be focusing on the total environment for the animal.

The Chairman indicated that the revised Council of Europe Convention represented minimal guidelines rather than regulations. He also indicated that justifications for not following the guidelines could be presented.

Examples of Issues/Questions Requiring Scientific Evidence

1. Rodents dislike wire floor.
2. Do rats have a need for gnawing, or is it an escape mechanism?
3. Is there an effect of noise (e.g., music) on animals?

Good Practice

1. It is difficult to base some aspects of good practice on specific factors other than experience. If we do not have enough data to change guidelines, we should ask the following:

2. Is it acceptable to use good practice?

3. Should the guidelines then remain the same?

COMMENTS FROM AUDIENCE

One participant indicated that we have not discussed work of experimenters who say we need to improve on the barren cages. Others indicated that, again, we need more data. We need to identify knowledge gaps and develop funding for experiments to address these gaps. In the meantime, there should be a moratorium on changes to the *Guide*. The idea "Let's do something" is a bad idea when there are no data to decide what to do. One person suggested that we continue using the current cage as we try to improve the environment.

After the two presentations that appear below (by Dr. Smith of work of Drs. Mabus and colleagues, and by Dr. Ottesen), the Breakout Group summarized as follows:

1. The diversity of guidelines, regulations, and traditions in different countries must be acknowledged. In publications, the set of guidelines/regulations under which the animal experimentation was conducted should be specified.

2. The treatment and conditions for laboratory animals should be evaluated regularly as new information becomes available. Animals have the same needs, wherever they live, and efforts must be made to conduct research in the best interest of the science and the animals.

3. It is premature to consider global standardization in the absence of scientific data.

4. Standards for maintaining animals should be posted, to eliminate the need for constant monitoring of research performed with animals. Flexibility to allow for professional judgment is critical.

5. There is a recognized element of fear of being forced into global standards, which implies a push toward increased regulation. Many feel the Canadian system is a better oversight model to emulate because of its flexibility.

6. There is concern about the acceptability of studies, particularly those that are based on animal preferences. Animal preferences do not always reflect the best welfare and scientific needs.

7. Because companies maintain research facilities in different countries, there is a need for harmonization, even in the absence of agreement. Although sensibilities differ among countries, it should be possible to reach some consensus.

Effects of Housing Density and Cage Type on Young Adult C57BL/6J Mice

Sarah L. Mabus, Abigail L. Smith, Jason D. Stockwell, and Cameron Muir

(Presenter: Abigail Smith)

The *Guide for the Care and Use of Laboratory Animals* (the *Guide*) (NRC 1996) specifies floor space requirements for laboratory mice of different weights. All cages must be at least 5 inches high. Floor space requirements (per mouse) are designated as at least 6 in^2 for mice less than 10 g, 8 in^2 for mice up to 15 g, 12 in^2 for mice up to 25 g, and more than 15 in^2 for mice weighing more than 25 g. The few peer-reviewed publications that address floor space needs of laboratory mice suggest that mice can be housed at densities higher than those recommended in the *Guide* and that mice housed at higher densities are healthier and less aggressive than mice housed at lower densities (Fullwood and others 1998; McGlone and others 2001; Van Loo and others 2001).

Rodent population densities have been shown to alter a number of normal and experimental parameters. In general, provision of less floor space either had no effect or was beneficial, resulting in enhanced immune responses and reduced mortality and aggression. Our study was designed to reveal how floor space and cage type might influence several parameters in young adult C57BL/6J (B6) male and female mice. The indices we studied were survival, aggressive behavior or injuries, body weight, food and water consumption, cage microenvironment (in-cage ammonia and CO_2 levels, temperature, and relative humidity), hair loss (a commonly observed characteristic in B6 mice, particularly females), urinary testosterone concentrations, and microscopic evidence of ammonia damage to nasal passages and eyeballs. We housed the mice in three readily available cage types that had different amounts of floor space. Populations of 4-week-old B6 mice were housed for 8 weeks in each cage type at four different densities—one compatible with recommendations in the *Guide* (approximately 12 in² per mouse) and three higher densities. A second, 4-week study was performed to determine whether recently weaned B6 male and female mice could be housed with even less floor space (reduced to 3.2 in² per mouse). We conclude that male and female B6 mice between the ages of 4 and 12 weeks can be housed with 5.6 in² of floor space per mouse without ill effect. This is approximately half the floor space recommended in the *Guide*.

METHODS

The cages we used were Thoren #1 ("shoebox," area = 67.6 in²) (Thoren Caging Systems, Inc., Hazleton, PA), Thoren #2 (weaning cage, area = 112.9 in²), and Thoren #3 (duplex, area = 51.7 in² per side). A total of 540 mice of each sex were included in the 8-week study, and 660 mice of each sex were included in the 4-week study. The densities are coded as follows: (1) = 12.9 in² per mouse; (2) = 8.6 in² per mouse; (3) = 6.6 in² per mouse; (4) = 5.6 in² per mouse; (5) = 4.5 in² per mouse; (6) = 3.8 in² per mouse; and (7) = 3.2 in² per mouse.

RESULTS

Eight-Week Study: C57BL/6J Mice House in Three Cage Types

Animal Health. All 1080 B6 mice that began the study survived, and we did not observe any aggressive behavior or injured mice. The mean weights of mice at the termination of this experiment were (± standard error [SE]) 20.4 ± 0.6 g and 29.8 ± 0.8 g for females and males, respectively. The incidence of alopecia among B6 female mice used in this study was relatively low, varying from 0 to 6% per treatment group, and was unrelated to cage type or housing density.

Microenvironment in Cages Housing C57BL/6J Male Mice. Ammonia levels were significantly affected by density. The levels at densities 3 and 4 significantly exceeded those in the two lowest densities and there was not a cage effect. In general, carbon dioxide concentrations increased with increasing densities. Mean CO_2 levels varied two-fold, ranging from 2733 to 5349 ppm, and were not affected by cage type. Temperature increased with increasing density, but mean temperatures varied ≤ 4°C and did not exceed the recommendation in the *Guide*. Mean relative humidity was reasonably constant across densities for each cage type.

Urinary Testosterone Concentrations for Male and Female C57BL/6J Mice.

Housing density had no effect on urinary testosterone levels of mice housed in any of the three cage types.

51.7-in² duplex cages. Mean female urinary testosterone levels (+ SE) significantly increased from 1.82 + 1.10 (baseline) to 4.35 + 2.37 ng/mg of creatinine by the end of the 8-week study, and male testosterone levels decreased (not significantly) from 2.86 + 1.16 to 1.87 + 0.91 ng/mg of creatinine by 8 weeks.

67.6-in² shoebox cages. Female mean urinary testosterone levels (± SE) increased from 1.29 + 0.86 ng/mg of creatinine (baseline) to 3.96 ± 1.22 ng/mg of creatinine (8 weeks), whereas male testosterone levels decreased from 3.19 ± 1.06 ng/mg creatinine (baseline) to 1.54 ± 0.71 ng/mg of creatinine (8 weeks). Both changes were statistically significant.

112.9-in² weaning cages. A single regression described the testosterone data from the weaning cages for all densities and both sexes. There was no significant difference in male or female mean urinary testosterone output between the baseline and 8-week samples. Unlike the results for males in 51.7 in² duplex or 67.6 in² shoebox cages, urinary testosterone levels increased for males between baseline and 8 weeks, although not significantly.

Four-Week Study: C57BL/6J Mice Housed at Higher Densities

Because we observed no deleterious effects of housing 20 C57BL/6J mice in 112.9 in² weaning cages for 8 weeks with 5.6 in² per mouse, we followed up with a 4-week study that evaluated the same parameters for mice provided with even less floor space—3.2 in² per mouse. We monitored the microenvironments of both male and female mice twice weekly in this study, and we also assessed both the noses and the eyeballs of selected mice microscopically at study termination.

Animal Health. Of the 1,320 mice that began this study, two mice were culled and one was found dead within the first 10 days of the study. These mice were replaced. No effect of density on the rate of weight gain was observed, and food and water consumption was not different among the densities. Density did affect the incidence of alopecia in female mice, which developed hair loss in one of six cages at density 4, two of six cages at densities 5 and 6, and five of six cages at density 7. Male B6 mice developed hair loss in one of six cages at each of the four densities.

Microenvironment in Cages Housing C57BL/6J Male or Female Mice. Mean ammonia concentrations (± SE) increased significantly with each increase in housing density (12.6 ± 1.1 ppm, 20.7 ± 1.1 ppm, 43.4 ± 1.1ppm, and 139.8 ± 1.1 ppm at densities 4, 5, 6, and 7, respectively). Nasal passages from selected mice in each density were examined microscopically for ammonia damage and were found to be normal. Eyes from 13 randomly chosen male mice were examined. The range of ammonia concentrations to which they had been subjected was 23 ppm to 410 ppm, including five mice from cages containing 198 to 399 ppm of ammonia. All of the examined eyes were histologically normal.

Carbon dioxide concentrations among the densities were not significantly different. Least squares means were greater for males (4,335 ppm) than females (3,103 ppm; $p < 0.05$). Least squares mean temperatures (± SE) were significantly higher for densities 5, 6, and 7 (28.3°C ± 0.2°C, 28.8°C ± 0.2°C, 28.4°C ± 0.2°C, respectively) compared with density 4 (26.6°C ± 0.2°C). The least squares mean relative humidity value (± SE) for the highest density (density 7: 57.1% ± 1.0%) was significantly higher than the two middle densities 5 and 6 (density 5: 53.0% ± 1.0%; density 6: 52.9% ± 1.0%) but not different from the lowest density (density 4: 56.3% ± 1.0%).

Urinary Testosterone Concentrations for Male and Female C57BL/6J Mice. Mean urinary testosterone levels were unrelated to housing density and were higher for males than females ($p = 0.002$). The mean baseline concentration for males was 14.4 + 4.1 ng/mg of creatinine, which increased to 26.8 + 12.6 ng/mg of creatinine by the end of the fourth week. For females, the baseline was 10.7 + 3.9 ng/mg of creatinine, which increased to 15.2 + 5.2 ng/mg of creatinine by week 4.

DISCUSSION

Based on gross measures, the health and well-being of the mice used in these studies were not affected by cage type or housing density. There were no significant differences among mice housed in three cage types, at any of the seven densities, in growth rates or food and water consumption. We did not observe aggressive or injurious behavior, and all mice survived the 8-week period of the first study. The incidence of alopecia among B6 female mice ranged from 0 to 6% in the 8-week study and was not associated with a particular cage type or housing density. The incidence of alopecia in the 4-week study was density dependent, with five of six cages containing affected female mice at the highest density.

In the 8-week study, in-cage CO_2 levels generally increased with density, and there were no apparent differences among the cage types—all reached maximum levels of approximately 5,000 ppm, the maximum allowable US workplace exposure limit during an 8-hour shift[a] or a 10-hour shift.[b] Increases in CO_2 concentration would be expected at higher

[a]http://www.osha.gov/dts/sltc/methods/inorganic/id172/id172.html
[b]http://www.cdc.gov/niosh/pel188/124-38.html

densities because CO_2 reflects the amount of respiration occurring within the cage. Mean temperatures generally increased with density, and on average, the difference between the mean high and low temperatures in each cage type was 2.5°C. The *Guide* indicates that temperatures ranging from 18 to 26°C are recommended for housing laboratory rodents. Exposure to a temperature higher than 29.4°C in animals not adapted to the high temperature could produce adverse clinical effects (NRC 1996). We have repeatedly measured in-cage temperatures at or exceeding 29°C and have not observed any adverse effects on multiple strains of mice. In this study, in-cage temperatures did not exceed 29°C. In-cage relative humidity was generally unaffected by cage density. The *Guide* indicates that relative humidity can vary widely, from 30 to 70%, and our results were well within that range.

In the 4-week study, average in-cage ammonia concentrations significantly exceeded 25 ppm at densities 6 and 7, reaching 43.4 and 139.8 ppm, respectively. The maximum allowable workplace ammonia exposure over an 8-hour period is 25 ppm.[c,d] Cages housing male mice had higher concentrations than those housing females. In-cage CO_2 concentrations were independent of density in the 4-week study. Temperatures were higher in cages housing the three highest densities but did not exceed 29°C. Humidity levels were variable and there was no clear relation to housing density.

Male urinary testosterone levels either remained relatively constant or decreased slightly over 8 weeks in the primary study. For female B6 mice housed in any of the three cage types, urinary testosterone levels increased over the course of the 8-week study. Irrespective of cage type, housing density did not influence urinary testosterone output of male or female B6 mice, although week and/or gender did. In the 4-week study, male urinary testosterone concentrations were uniformly higher than female concentrations. As found in the 8-week study, neither housing density nor cage type influenced hormone concentrations. However, in contrast to the 8-week study, male hormone levels increased over the 4-week period. Female hormone levels increased in both studies. It may be noted that the testosterone concentrations in the 4-week study were substantially higher than those in the 8-week study. This interassay variation is expected and makes it essential that values to be directly compared must be the result of simultaneous assays.

Although the OSHA and NIOSH standards cited above indicate that workplace exposure to ammonia should not exceed 25 ppm over 8 hours

---

[c]http://www.osha.gov/dts/sltc/methods/inorganic/id188/id188.html
[d]http://www.cdc.gov/niosh/pel88/7664-41.html

or 35 ppm over a 15-minute period, two factors can substantially reduce human exposure in animal facilities. First, when filter tops are removed from rodent cages, there is an immediate dilution effect by mixing with ambient air. Second, as is the case with Mus m 1 allergen exposure in mouse rooms (Schweitzer and others 2003), exposure can be greatly reduced by husbanding rodents on ventilated tables. In addition, the type of bedding that is used to house the mice can have a significant impact on in-cage ammonia concentrations (E. Smith and others, submitted). The finding that rabbits exposed continuously to > 400 ppm had opacities greater than one-fourth to one-half of their corneas (Coon and others 1970) was the basis for our microscopic examination of mouse eyes in the 4-week study, and lesions were not observed.

Results of our 8-week study indicate that only the least squares mean ammonia level (49 ppm) in density 4 in the 112.9-in^2 weaning cages exceeded the concentration considered unhealthy for humans. This result may have been anomalous because the mean concentration for that density in the second study was 20.1 ppm. However, in the 4-week study, ammonia levels were very high (43.4 and 139.8 ppm) in cages housing mice with less than 4.5 in^2 of floor space. None of the mice in either of the two studies showed evidence of ammonia toxicity, despite exposure to > 200 ppm in some individual cages in the 4-week study. Nonetheless, given the OSHA workplace standards for humans of 25 ppm[c,d] (Vigliani and Zurlo 1956), the use of ventilated changing tables should be encouraged in mouse rooms.

The floor space recommended in the *Guide* was based on best professional judgment at a time when there was very little peer-reviewed literature on the topic. We are attempting to apply scientific methods to learn the real floor space needs of mice. Thus, we recommend housing C57BL/6J male and female mice, aged 4 to 12 weeks, in cages that provide not less than 5.6 in^2 of floor space per mouse. This housing translates to nine mice per side of duplex (51.7 in^2) cages, 12 mice per shoebox (67.6 in^2) cage, and 20 mice per weaning (112.9 in^2) cage.

Our results and those of others (Fulwood and others 1998; McGlone and others 2001; Van Loo and others 2001) have consistently pointed to the same conclusion: Mice that are housed at higher densities tend to be healthier and less aggressive toward their cage mates. For this reason, it is necessary to re-evaluate the current guidelines in the context of what is known about this social species. Animal care should not be dictated by the anthropomorphic perceptions of animal caretakers and regulatory

---

[c]http://www.osha.gov/dts/sltc/methods/inorganic/id188/id188.html
[d]http://www.cdc.gov/niosh/pel88/7664-41.html

bodies. Animal care staff almost always want to do what is best for the animals, but they may need to be educated in the area of rodent housing density. The role of the *Guide* is to ensure that laboratory animals are well treated and housed in a species-appropriate manner.

This housing density study and others reported in the literature have included only a few inbred mouse strains. Universal provision of the floor space needs of mice may be difficult, and strain variation is to be expected. We are currently using the same protocols described herein to evaluate the needs of young adult BALB/cJ, NOD/LtJ, and FVB/NJ mice; and we have data indicating that there are, indeed, differences (A.L. Smith, manuscript in preparation).

ACKNOWLEDGMENTS

We thank Drs. Richard Smith and Ralph Bunte for microscopic evaluation of eyes and nasal passages, respectively, and Drs. Beverly Paigen and David Threadgill for helpful comments during manuscript preparation. This work was supported by grant RR12552 from the National Center for Research Resources, National Institutes of Health; by a Henry and Lois Foster Foundation grant (ACLAM Foundation); and by funds from the Jackson Laboratory.

REFERENCES

Coon, R.A., Jones, R.A., Jenkins, L.J., Siegel, J. 1970. Animal inhalation studies on ammonia, ethylene glycol, formaldehyde, dimethylamine and ethanol. Toxicol Appl Pharmacol 16:646-655.

Fullwood, S., Hicks, T.A., Brown, J.C., Norman, R.L., McGlone, J.J. 1998. Floor space needs for laboratory mice: C57BL/6 males in solid-bottom cages with bedding. ILAR J 39:29-36.

McGlone, J.J., Anderson, D.L., Norman, R.L. 2001. Floor space needs for laboratory mice: BALB/cJ males or females in solid-bottom cages with bedding. Contemp Top Lab Anim Sci 40:21-25.

NRC [National Research Council]. 1996. Guide for the Care and Use of Laboratory Animals. Washington, DC: National Academy Press.

Schweitzer, I.B., Smith, E., Harrison, D.J., Myers, D.D., Eggleston, P.A., Stockwell, J.D., Paigen, B., Smith, A.L. 2003. Reducing exposure to laboratory animal allergens. Comp Med 53:487-492.

Van Loo, P.L.P., Mol, J.A., Koolhaas, J.M., Van Zutphen, B.F.M., Baumans, V. 2001. Modulation of aggression in male mice: Influence of group size and cage size. Physiol Behav 72:675-683.

Vigliani, E.C., and Zurlo, N. 1956. Experiences of the Clinical Del Lavoro with maximum allowable concentrations of industrial poisons. AMA Arch Ind Hlth 13:403.

New Housing Standards for Rats and Mice Developed with Focus on the Needs of the Animals

Jan L. Ottesen

INTRODUCTION

The Expert Group on Rodents and Rabbits of the Council of Europe has clearly stated that its objective is to meet the needs of animals. As part of the extensive background information in the Preamble to their proposal for revision of Appendix A of the European Convention ETS 123, the following statements appear (Stauffacher et al. 2002):

The exact numeric values for minimum cage sizes and heights as well as for maximum stocking densities can never be scientifically evaluated and "proved." Working out minimum requirements with respect to animal welfare and to supposed well-being of laboratory animals is a political question. Nevertheless, the decision-making process should be based first and foremost on sound arguments on the biology of species and strains in question. During discussion it should be carefully distinguished between biological facts, scientific evidence and practical experience on one side and ethical principles of animal protection and the assessment of economical and political reason on the other side.

It is important to bear this position in mind during any discussion on "Science-based Guidelines for Laboratory Animal Care."

In 1999, at the Third World Congress on Alternatives and Animal Use in Bologna, Coenraad Hendriksen (2000) proposed the "Three C Principles"—Common sense, Commitment, and Communication. Dr. Hendriksen described these principles as drivers toward implementation of the principle of the "Three Rs"—Refinement, Reduction, and Replacement (Russell and Burch 1959).

Of course, optimal guidelines for laboratory animal care should be science based. However, scientific proof is often not possible to obtain (e.g., determination of the exact numeric values for minimum cage sizes and heights). Furthermore, scientific proof is often used defensively as a prerequisite before introducing new environmental enrichment ideas, which unfortunately often stops further progress. At Novo Nordisk, we have tried to use common sense when we develop new housing facilities for our experimental animals, based on expert views on animal needs. We do not yet have scientific proof that all of our initiatives have resulted in increased animal welfare, yet one could argue that neither do we have proof that demonstrates the opposite. In the absence of proof, we provide what common sense tells us is good animal welfare.

Environmental enrichment is one of the major ways of trying to improve the welfare of laboratory animals in our care. Freedom of movement and a structured environment that allows natural behavioral patterns of the animals are considered an enrichment of the environment. However, rats in particular are highly adaptive, and it is difficult to prove that environmental enrichment does in fact increase the welfare of these animals. It seems obvious from studies both of rats in captivity and of laboratory rats released to semi-natural conditions (Berdoy 2002) that they rear to perform grooming and to look out, as part of their natural behavior. In our new rat cage system, which has been implemented during the testing period, we have increased the height of the cage from 18 to 30 cm. In the future, we plan to study the potential benefit to the rats of the new housing conditions (see below). In the meantime, we believe that the welfare of the rats is not jeopardized under current housing conditions.

FROM GUIDELINES TO LAW

In 1998, while revising the 1986 European Convention for the Protection of Vertebrate Animals Used for Experimental and Other Scientific Purposes (ETS 123 Guidelines), the Council of Europe established a number of working groups to review the different animal species used as experimental animals. Based on proposals from these expert groups, species-specific sections have been prepared. In addition, extensive background information containing scientific evidence as well as practical experience has been compiled to support the expert groups' proposals. For most of the species, the proposals for species-specific provisions have been finalized, although officially, they are still considered draft versions, not currently in effect (www.coe.int/animalwelfare).

In Denmark, the government decided not to await the final revision of ETS 123. In August 2003, the Ministry of Justice issued a government order that authorized all of the above-mentioned "finalized" draft versions, thereby changing the content from guidelines to law. It should be noted, however, that any necessary major changes in building constructions have until 2007 to be in place.

NEEDS OF RATS AND MICE

At the end of 1999, in an effort to identify and establish the most important needs of mice, rats, guinea pigs, rabbits, and dogs in their natural habitat, Novo Nordisk A/S and the Danish Animal Welfare Society invited internationally recognized animal welfare experts to participate in several workshops. Based on the results from those expert workshops, new laboratory animal housing prototypes have been developed that take the needs of animals into consideration much more than in prior years (Ottesen et al. 2004).

Mice

Mice are social animals that prefer to be with conspecifics. They have a need to live in stable, harmonious groups, although it might be necessary to separate adult male mice to avoid their aggressiveness. Mice need to be able to rest, hide, and build nests. They also have a need for complex and challenging surroundings. Mice are nocturnal animals, and therefore need darkness.

At Novo Nordisk, we believe that the needs of the mice can be accommodated for the most part in types III and IV macrolon cages (800 and 1800 cm^2, respectively). The optimal cage size will depend on the weight of the mice, the group size, and the extent of environmental enrichment. Further improvement of the cage may be accomplished by using a commercially available lid that is 7 cm higher.

Rats

Rats are social animals that need a structured and enriched environment with access to both hiding and viewing places. They need space for rearing, climbing, gnawing, digging, and grooming.

A few years ago, Novo Nordisk implemented the European type IV macrolon cage system (a cage with 1800 cm^2 of floor space) as the standard cage system for rats. This cage has been enriched further in the replacement of traditional lids with specially designed "elevated lids," increasing the height of the cages from 18 to 30 cm. This replacement affords the rats the possibility of visual control of the environment and space for rearing and grooming. A shelf in the cage provides the rats with additional possibilities for exploring, exercising, jumping, and looking out, as well as improved hiding possibilities (Figure 1).

DISCUSSION

Economy is sometimes used as an argument against larger cage sizes for experimental animals, and various calculations are used to support the argument. The annual total budget for the animal facility is often used as reference. Even though the necessary investment in new cage systems is 25, 50, or 100% of the facility's annual budget, it should be noted that this expense is a one-time investment. Compared with the annual budget of the company or institution, that expense most likely totals a percentage that is less than one digit.

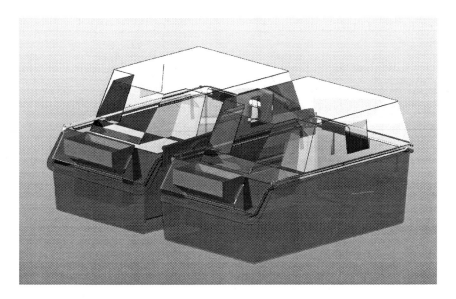

FIGURE 1 (A) The European type IV macrolon cage (1800 cm^2) is used for housing rats. A new, high lid increases the total height of the cage from 18 cm to approximately 30 cm, which allows the rats to rear and perform grooming. The possibility of connecting two or more cages with a tunnel has been developed and will allow larger group sizes or more room for the rats. (Figure 1A: Courtesy of Scanbur BK.)

FIGURE 1 (B) Provision of a shelf gives the rats more choices and provides both hiding and nesting opportunities (nesting material partly removed on photograph).

At Novo Nordisk, we believe that our new cage systems for mice and rats will cover their basic needs much better than the smaller, traditional cage systems. Beyond ensuring the conditions we believe greatly improve the welfare of the animals, the new system appears much more pleasant and inviting and should therefore contribute to a broader acceptance of experimental animal use by the public.

It will be necessary for laboratory animal care guidelines to deal with minimum requirements. It is acknowledged that for some rodent studies (e.g., in research on brain development), less enriched cages may be required. For other studies (e.g., some feeding studies), single housing of the animals may be necessary. Nevertheless, using as examples the studies that require less enrichment is not a valid argument for housing all animals under minimal conditions.

If possible, housing standards should be science based; however if many parameters need to be scientifically proven to implement new housing standards, common sense should also be applied.

ACKNOWLEDGMENTS

Thanks are due the animal technicians in the Novo Nordisk Environmental Enrichment Committees for the commitment demonstrated in the excellent translation of expert advice from M. Stauffacher, R. Hubrecht, R. Murison, P. Hawkins, A. Kornerup Hansen, V. Baumans, M. Ritskes-Hoitinga, I.M. Jegstrup, and L. Lidfors into new housing systems for laboratory animals.

REFERENCES

Berdoy, M. 2002. The Laboratory Rat: A Natural History (Video/DVD). London: Oxford University.

Hendriksen, C.F.M. 2000. Replacement, reduction and refinement and biologicals: About facts, fiction and frustration. In: Balls, M., van Zeller, A.-M., Halder, M.E., eds. Progress in the Reduction, Refinement and Replacement of Animal Experimentation. Amsterdam: Elsevier Science B.V. p. 51-63.

Ottesen, J.L., Weber, A., Gürtler, H., Friis Mikkelsen, L. 2004. New housing conditions: Improving the welfare of experimental animals. ATLA 32(Suppl 1) (In Press).

Russell, W.M.S., and Burch, R.L. 1959. The Principles of Humane Experimental Technique. London: Methuen & Co. Ltd. p. 1-238.

Stauffacher, M., Peters, A., Jennings, M., Hubrecht, R.C., Holgate, B., Francis, R., Elliott, H., Baumans, V., Hansen, A.K. 2002. Future principles for housing and care of laboratory rodents and rabbits. Report for the Revision of the Council of Europe Convention ETS 123 Appendix A for Rodents and Rabbits. PART B. (http://www.coe.int/T/E/Legal_affairs/Legal_co-operation/Biological_safety,_use_of_animals/Laboratory_animals/GT123(2001)4%20Final%20PART%20B%20Rodents.pdf). Strasbourg: Council of Europe.

Breakout Session: Approaches for Implementing Current US and European Guidelines for Housing Standards for Dogs and Cats

Leader: Robert Hubrecht

Rapporteur: Thomas Wolfle

The session began with a review of dog cage or pen size guidelines, policies, and regulations in the United Kingdom and the Council of Europe (CoE 2001), in Canada (according to pertinent documents of the Canadian Council on Animal Care [CCAC; www://ccac.ca]), and in the United States (according to the *Guide for the Care and Use of Laboratory Animals* [NRC 1996]). Video examples were shown of stereotypies in dogs (e.g., cage chewing) and remediation with cage enrichment. A time-budget chart was presented showing that dogs spend large amounts of time on elevated resting platforms. Early socialization, habituation, and training were reviewed, and a behavioral technician playing with dogs in an enriched play environment was presented to set the stage for discussing space requirements.

Participants expressed the concern that too little space restricts group size and associated social interaction. They felt that the size of cages or pens should be judged adequate only when the following needs are accommodated: (1) species-specific activity and interaction; (2) enrichment, such that the animals are able to manipulate and control particular aspects of the environment; and (3) essential space for resting, temperature control, sanitation, and noise control. To discuss these concerns, the leader posed the following questions:

What factors led to the CoE's large cage size?

Participants noted that smaller cages are required in the United States compared with those described in the CoE documents, yet few abnormalities are noted in the US cages when adequate socialization is provided. Participants seemed to agree that science-driven cage size recommendations are needed, and that ever-larger cage size mandates without supportable science are not justified. Focusing on cage size, rather than on behavior, forces the use of engineering, rather than on performance, standards.

How should environments be designed?

Participants indicated that performance-oriented approaches to cage size and environments are the most appropriate. The definition of these performance goals revolves around behavioral assessment and includes the considerations described below.

Some participants felt that spatial requirements for dogs in research should take into account the whole experimental program as well as species needs (although dogs are not different from other species in this respect). In addition, they felt that acceptable environments should allow the following: positive interaction with humans, harmonious conspecific social housing and exercise, and opportunities for reasonable species-specific behavior such as play and gnawing. Moreover, it was felt that the occurrence of stereotypies and other abnormal behavior should be minimized. The life-to-death experience was considered by some participants to be critical in ensuring high standards of welfare and high-quality science. One participant felt that standards at supply sources (i.e., breeders) are important, and there should be good communication between suppliers and users. Consideration should be given to the animals' use in acute versus long-term studies, or survival studies in which the animal might be adoptable. Some participants felt that housing standards should take into account the adaptability of the species (although it is not clear that this factor is any greater than for some other species commonly used) and the variability of different breeds.

Should the length of time that a dog spends in a facility be a factor in the standards provided?

The focus of the question above was whether additional attention should be paid to addressing the needs of dogs used in longer-term studies. Some participants argued that dog housing should meet the needs of the species regardless of the length of time the animals might need to be housed in it. In addition, some opined that a multiplicity of standards for studies of different length might lead to unnecessary bureaucracy and confusion. The participants did not reach any consensus on this question.

Should standards be based on engineering, performance, or a mixture of the two?

Participants expressed the belief that a mixture of the two standards is best. There is a need to concentrate on performance standards because they often indicate the true success or nonsuccess of the dog enclosure. However, engineering standards are also useful to ensure the fulfillment of minimum standards.

How should changes be implemented?

When developing standards, participants indicated that it is necessary to begin with the requirements of the dog, and then move on to regulatory issues. More guidance is needed to evaluate the adequacy of environments in order to assess performance standards. It is important that any new standards be phased in, and that anticipated costs of implementing new standards be included in applications for funding. Many felt that more training of personnel is needed to recognize normal and abnormal animal behavior. Participants emphasized the need for consideration of the research mission in making changes. Involvement of the Principal Investigator in planning prospective changes in housing and enrichment is essential to ensure the high quality of ongoing data.

What arguments should be used in the process?

Many participants felt that changes should be made with due caution and based upon scientific evidence, professional judgment, and widely accepted "Best Practice." To avoid reinventing the wheel, and to help harmonize international standards, it is advisable to refer to existing standards (e.g., NRC 1996, CoE 2001, and other nations' codes of practice).

What does the public expect?

The view was expressed that science is carried out on the public's behalf and indirectly with their consent in general terms. Therefore, the public has a particular concern for the welfare of species commonly used as companion animals, and standards used in the laboratory should reflect that fact.

How should economic arguments be weighed against biological arguments?

This question, stated another way, asks how the cost-benefit of animal research should be established. Some expressed the view that political decisions mandating engineering standards are likely to be unduly expensive without concomitant benefit to the animals.

Good science and good welfare go together, and ongoing assessments provide valuable answers to the cost-benefit question. Central to the assessment is an understanding and application of science-driven performance standards. While many felt that global harmonization of animal care and use practices offers many potential benefits both to humans and to animals, some expressed the notion that harmonization would be unlikely if the engineering standards are politically motivated.

Is further science needed? How should it be directed?

Participants readily endorsed the need for additional research (such as is listed below) for better planning of research. However, some felt that greater use of sound experimental design and statistics is necessary to accommodate any new science-derived changes in the use of animals in research. Participants also recommended expanded use of biologic telemetry.

What are the needs for housing- and welfare-related research?

- Areas described above.
- Further investigation/research is needed regarding economical and practical ways of enriching the pen environment and of taking into account the needs and sensory modalities of dogs.
- Relation between pen size, contents/structures/other enrichment, number of individuals, and behavior, preferably under carefully controlled experimental conditions.
- Ways to ameliorate the negative effects of single housing. Do exercise plans for single-housed dogs actually make a difference?
- Ways to prevent and manage aggression.
- Determination of the effects of sound on dogs
- Cost versus benefit of different toys and chews (no such studies have been attempted in canids).
- Comparison of different methods of presentation of toys and chews, and determination of the effectiveness of various types of enrichment in single housing and in larger groups.
- Influence of breeding for the selection of desirable characteristics.
- Psychological and physiological effects of transport.
- Design of metabolism cages to reduce their impact on dog welfare.

REFERENCES

CoE [Council of Europe]. 2001. European Convention for the Protection of Vertebrate Animals Used for Experimental and Other Scientific Purposes (ETS 123). Future Principles for Housing and Care of Laboratory Rodents and Rabbits. Strasbourg: Council of Europe.

NRC [National Research Council]. 1996. Guide for the Care and Use of Laboratory Animals. 7th ed. Washington, DC: National Academy Press.

Breakout Session: Nonhuman Primates

Leader: David Whittaker

Rapporteur: Randall J. Nelson

Participants discussed the questions that appear below, in general consideration of the guideline revision process—Who, How, and Outcomes:

How should the next revision of the *Guide for the Care and Use of Laboratory Animals* (the *Guide*) (NRC 1996) be conducted?

Participants believed that expert groups chosen to address specific issues should conduct the revision of the *Guide.*

Is the lack of scientific knowledge in an area sufficient reason not to move forward with revision?

Session Leader David Whittaker believed not.

The Council of Europe (CoE) formed expert groups to deal with formulated guidelines, but eventually legislative recommendations for acceptance or nonacceptance of guidelines developed. The CoE participants believed that competent authorities (i.e., ministries) who implemented laws protected against conflict of interest.

Is a smaller group of experts more efficient in developing guidelines?

Participants affirmed small-group efficiency in contrast to the inertia of larger groups. They also opined that competent authorities should agree in advance to abide by the recommendations of the expert groups unless they vary radically from socially accepted norms. Moreover, they felt that: (1) expert groups should provide technical information early in the process because subsequent change is difficult to implement; (2) large groups are more costly and difficult to manage; and (3) industry should be involved from the beginning, as should all of the stakeholders.

To what extent has social housing of nonhuman primates (NHPs) been accomplished in Europe?

Individuals in the group indicated that only about 1% of NHPs in the United Kingdom are singly housed and in those cases only for scientific reasons. Many felt that positive reinforcement training is beneficial in facilitating the handling of socially housed NHPs. Some also felt that regulations should influence, rather than require, compliance with factors such as social housing and that influence should be exerted to achieve "best/good practices."

How does one determine best/good practices?

There was no consensus among the participants on this question. It was pointed out that the United Kingdom maintains a central clearinghouse for best/good practices However, the UK does not promote the blanket utilization of justification of exceptions because doing so discourages the consideration of alternatives and refinements. Nevertheless, participants felt that there should be ready access to information about best/good practices so that refinements can be made with a minimum of regulatory burden.

Further discussion elicited the following opinions from the participants:

- Consistency in the guidelines and the authority to impose them is lacking in instances in which few scientific studies are available to substantiate expert opinion and professional judgment.
- The scientific and animal care communities need to convince competent authorities to be supportive of the need to gain more scientific data on factors such as cage sizes. They also need to convince society that such studies are worth the initial investment, because it may be perceived that funds are being redirected from health-related research.
- The fundamental issues are economics and politics.
- In studies that will have an impact on welfare issues, expert groups should agree on the range of experimental variables before the studies are performed, to avoid instances in which the scientific validity of the results is called into question.

How do we legitimize the science needed to fill gaps in the literature related to welfare issues?

Participants provided perspectives and outlined the following potential strategies for change:

- Scientists and veterinarians need to be proactive with legislators from the very beginning to effect the change in societal attitudes needed to make funding of these studies more likely. The Medical Research Council gives monies for appropriate changes in approaches to be made, but if they give monies, the resultant changes are required. Thus, they have the "force of law" behind their support.
- Data mining may be beneficial in obtaining needed scientific data with little or no cost. The data may already be available in some instances.
- Veterinary outreach to investigators and to the community is an important way to educate others about best/good practices.
- Qualified experts should attempt to identify the "bad science" in extant guideline documents, thus increasing the validity and applicability of the documents before refining them or rewriting them.
- Participants recommend that a list of perceived gaps in the scientific basis for welfare decisions should be maintained. Although not all gaps may be filled at the time of any revision of guidelines, maintenance of these lists will facilitate their consideration at a later date. These lists act as bellwethers for areas where additional guidance may be needed. Additional indicators may come from indirect observations. For example, in instances in which guidance is less than adequate, interinstitutional variation in the implementations of guidelines due to professional judgment may indicate areas where additional guidelines are necessary to promote consistency in welfare and care.
- Minimum acceptable standards may need to be established to facilitate consistency in enforcement. Without minimum standards, enforcement may be perceived as arbitrary.
- Difficulties may arise when members of expert groups are included for political reasons. All stakeholders should be included, but representation should be balanced to ensure efficiency.
- Some participants asked whether the questions being asked are the right ones. It was suggested by some that the goal should be the maximum improvement in welfare relative to the amount of effort generated to reach that goal.

What concerns exist relative to the way the *Guide* deals with NHPs?
Participants identified four concerns:

- Occupational Health—Occupational health is fraught with variability across institutions as a result of vagaries in guidelines. Personal protective equipment in laboratories and proximity issues were discussed. Exposure as a function of proximity to NHPs and duration of exposure should be dealt with more specifically because investigators are looking for guidance.

- Positive Reinforcement Training—Positive reinforcement training is not dealt with in detail but could facilitate welfare in instances of social housing.
- Social Housing—Species-specific considerations are not extensive in the *Guide*. Social contact without social housing (by touch windows) may allow animals to withdraw when necessary, which achieves welfare goals. However, it may also reduce the vulnerability of individuals to injury, which is of concern to those who question the utility of social housing as a default condition.
- Animal Welfare—There is a need to think about welfare from the "standpoint of the animal."

CONCLUSION

The participants stressed that "one size does not fit all," especially with respect to NHPs. Individuals of the same species often behave quite differently under the same environmental and behavioral situations. Participants felt that the "Redbook" (NRC 2003) successfully maintains this philosophy throughout discussions of individual experimental situations and other documents should be continued in this stance. It was felt that guidelines should include the consideration of an individual's needs, experimental contingencies, and ethical responsibilities.

REFERENCES

NRC [National Research Council]. 1996. Guide for the Care and Use of Laboratory Animals. 7th Ed. Washington, DC. National Academy Press.

NRC. 2003. Guidelines for the Care and Use of Mammals in Neuroscience and Behavioral Research. Washington, DC. The National Academies Press.

Breakout Session: Rabbit Housing

Leader: Vera Baumans

Rapporteur: Jennifer Obernier

Guidelines for this breakout session encouraged its leader and participants to debate current research on rabbit housing standards and guidelines in light of current scientific information. In particular, the participants discussed the pros and cons of group housing, the standards and guidelines that govern minimum caging size, and the climate of the housing environment.

GROUP HOUSING

Both the Council of Europe (CoE 1986) and the *Guide for the Care and Use of Laboratory Animals* (the *Guide*) (NRC 1996) recognize that rabbits are social creatures and should be housed, when possible, in social groups to maximize species-specific behaviors and minimize stress-induced behaviors. The participants readily agreed that in some cases, for scientific or veterinary reasons, rabbits should not be group housed. However in most cases, social housing is an excellent idea, provided there is complexity in the caging. Complexity may include providing visual barriers and hiding places to minimize aggressive encounters and to allow animals to avoid contact by withdrawal. Social housing does have drawbacks, such as fighting, which can cause injury, and the need for improved animal husbandry and housing, to ensure the adequacy of food and water and the hygiene of the cage.

CAGE SIZE

Participants discussed the Council of Europe proposal to revise Appendix A of the European Convention ETS 123. This revision establishes a new minimum cage size standard based on weight, such that one or two compatabile rabbits of less than 4 kg should be housed in a cage with a floor area of 42 cm² and a height of 45 cm, including a raised area of approximately 55 × 30 cm. Without a raised floor space, the floor area for one rabbit should be 5600 cm² and 6700 cm² for two rabbits. Existing standards and guidelines for rabbit caging size also base minimum cage size on weight, including (1) the European Convention ETS 123, which sets a standard of 2500 cm² floor area and 35 cm of vertical space for a rabbit of less than 4 kg, and (2) the *Guide*, which recommends allocating a rabbit less than 4 kg approximately 2800 cm² of floor space and approximately 35.5 cm of vertical space. During the discussion, participants agreed that basing cage size on weight is not optimal because it does not take into consideration that young rabbits are active and need more space in relation to their body weight than adults.

CLIMATE

The Council of Europe has proposed specific standards for the climate of rabbit housing, including temperature and humidity standards. However, participants pointed out that these standards were arbitrarily set and not based on science. Furthermore, it is unclear whether fluctuations above and below these standards during cleaning and scientific manipulation have any impact on animal well-being. Some participants felt that additional research is needed to understand seasonal fluctuations and the effects of extending the boundaries of the temperature and humidity standards proposed by the Council of Europe.

REFERENCES

CoE [Council of Europe]. 1986. European Convention for the Protection of Vertebrate Animals Used for Experimental and Other Scientific Purposes (ETS 123). Strasbourg: Council of Europe.

NRC [National Research Council]. 1996. Guide for the Care and Use of Laboratory Animals. 7th Ed. Washington, DC: National Academy Press.

Session 4

Environmental Control for Animal Housing

Environmental Controls (US Guidance)

Bernard Blazewicz and Dan Frasier

CURRENT US GUIDANCE

Current guidance regarding environmental conditions for vivariums is primarily found in industry and government publications. The most widely accepted publication and the primary reference on animal care and use is the *Guide for the Care and Use of Laboratory Animals* (the *Guide*), published by the National Research Council (NRC 1996). Other pertinent references include the American Society of Heating, Refrigeration, and Air-Conditioning Engineers (ASHRAE 2003), the National Institutes of Health Design Policy and Guidelines (NIH 1999), the Biosafety in Microbiological and Biomedical Laboratories (CDC/NIH 1999), and the US Department of Agriculture ARS 242.1M (USDA 2002).

The *Guide* places emphasis on performance standards, as opposed to engineering standards, for environmental control. Performance standards are viewed to be more flexible and more concerned with the outcomes than engineering criteria.

To apply the *Guide* effectively, a team approach is recommended whereby facility users and designers can share expertise to meet desired outcomes. The *Guide* is not a how-to-build handbook on vivarium design; it provides broad recommendations for environmental conditions that have proven to work well. Individuals responsible for well-designed facilities begin with a thorough understanding of the scientific needs, and then translate that information into a facility that meets the expectations of the users.

The *Guide* allows for interpretation or modification in the event that acceptable alternative methods are available, or unusual circumstances arise when deviating from the *Guide*. For example, ventilation rates that exceed 10 to 15 air changes per hour (ac/h) would be allowable, given appropriate justification. When deviating from the *Guide*, thought should be given to other environmental factors that may be affected by the deviation. In the case of air change rates, it is possible that air movement, diffusion pattern influence on the animal's microenvironment, and the relation of the type and location of supply-air diffusers and exhaust vents would warrant further consideration.

ENVIRONMENTAL CRITERIA

Environmental criteria topics that have been discussed include the following: temperature and humidity, ventilation rate, lighting, containment, and air quality. Each of these topics is briefly described below.

Temperature and Humidity

The most common source of data for temperature and humidity is ASHRAE; however, most data are outdated and date back to the 1950s or 1960s. Some researchers believe that the measurements concluded from past heat and moisture data are too low for today's animals. Recent rodent data have provided evidence that rodents have higher metabolisms and heat generation (Riskowski and Mermazedeh 2000).

Ventilation Rate

Ventilation rates have historically followed the 10 to 15 ac/h (fresh air) recommendation from the *Guide*. This range has proven to be a good range although different approaches allow lower ventilation rates while maintaining a stable animal room environment (i.e., ventilated caging systems). Some applications, species, and rooms require more than 15 ac/h. It should be emphasized that 10 to 15 ac/h has historically proven successful in managing most animal thermal and respiration loads and equipment loads. However, the *Guide* is clear that calculations must be performed to determine the air change rates required to remove the thermal and moisture loads and provide any additional make-up air exhaust devices (i.e., fume hoods or biosafety cabinets).

Lighting

Lighting normally consists of dual levels (day/night) and override for cleaning. Present methods of monitoring and controlling lighting are to use the building automation or environmental monitoring systems. Typical ranges applied are 30 foot-candles (f.c.) for day, 0 f.c. for night. Lighting levels for cleaning range from 70 to 100 f.c. for 1 hour.

Containment

Reduction of cross-contamination between holding rooms is normally accomplished through pressurization—supply/exhausting air to/from the room to direct air in or out of the room. Quarantine, isolation, biohazards, and nonhuman primates should be kept under negative pressure. Pathogen-free animals, surgery, and cleaning and equipment storage should be kept under positive pressure. The bubble diagram in Figure 1 is an illustration of different types of pressure schemes that can be found in a vivarium.

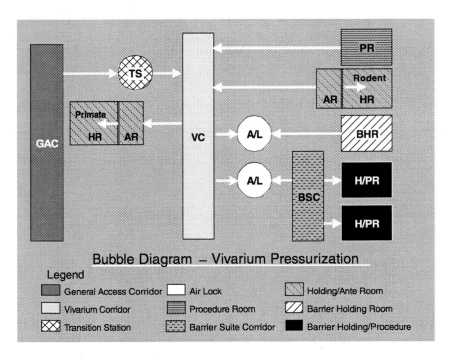

FIGURE 1 Example of the different types of pressure schemes in an animal research laboratory vivarium.

Air Quality

Current guidelines provide no criteria to judge air quality. Past practices have included the use of high-efficiency particulate air (HEPA) filters. The *Guide* recommends HEPA filters for certain areas—surgery and postoperative holding rooms. Heating, ventilation, and air-conditioning (HVAC) systems normally use ASHRAE-rated filters, which are effective at keeping HVAC system components clean and extend the life of HEPA filters.

TECHNOLOGICAL ADVANCES

Recently, a greater focus has been placed on the room environment, which includes room allergen levels, the migration of airborne pathogens, temperature/humidity comfort levels, and biosafety containment. Animal facilities are now utilizing a more comprehensive and scientific approach to address these concerns. The analytical tool of choice to aid in the design of these rooms is computational fluid dynamics (CFD). CFD has been used successfully over the past 20 years for accurate modeling of air currents, temperature, and humidity levels. The method is further evolving to include fresh air dwell times, particulate movement, stagnation, and projected odor levels. Other parameters that may be studied include inlet diffuser type, animal heat loads, cage/rack placement, and exhaust air systems placement. CFD provides a visual representation of the effects of airflow in the holding room and a better understanding of the room dynamics. Together, these advances provide better scientific data for the development of future guidelines. Figure 2 is an illustration of a sample of CFD output that was used to determine odor migration in a canine holding room, modeling several different versions of supply/exhaust placement, to determine which arrangement provided better containment of odor (AALAS 2003).

GAPS IN CURRENT GUIDANCE AND CRITERIA

Noise and Vibration

Currently, there is no acoustical criterion for animal rooms contained in the *Guide* or from ASHRAE. The hearing ranges of animals are different from humans, and the ranges are different among species. Examples of ranges are shown in Figure 3.

Limited published data are available on sound sources and mitigation techniques. Numerous internal studies have been performed, and techniques and strategies have been developed to mitigate noise, which

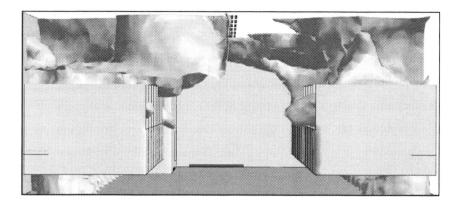

FIGURE 2 Computational fluid dynamics (CFD) analysis of a canine holding room. CFD was used to develop a three-dimensional model of a gas concentration in a room at the prescribed concentration level of 5 ppm. End view, NH_3 isosurfaces measuring 5 ppm.

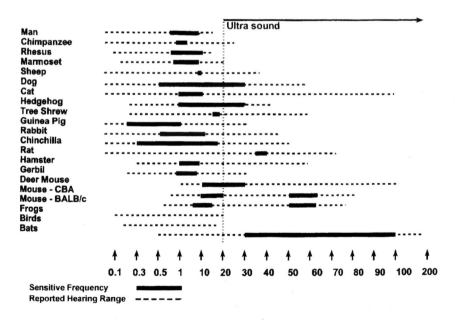

FIGURE 3 Examples of differences among the hearing ranges of humans and various animal species. Modified from Warfield (1973) and Sales and Pye (1974).

can be helpful in developing criteria for future updates to current guidelines. Published vibration criteria for animal facilities are also very limited, but again, numerous internal studies have been performed that could help the industry establish such criteria.

Performance Standard for Ventilated Cages

Ventilated caging systems have evolved into many different airflow strategies based on the work of various manufacturers. Generally, manufacturers have worked closely with animal research professionals to develop caging systems that have well-founded concepts. It is recommended that the scientific community, along with industry professionals and manufacturers, develop a performance standard for ventilated cages to identify the knowledge base and the most important criteria.

SUMMARY

• Established guidelines have proven to work well but have not been updated to reflect new trends in vivarium research that affect the environment.
• Technology has advanced our understanding of the macro- and micro-environment.
• Independent research and testing have produced new insights that have affected the vivarium environment.
• Additional guidance and work are required to close the gaps.
• Variances based on scientific data are recognized and allowed in the *Guide*.

RECOMMENDATIONS

• Update ASHRAE guidance based on current research.
• Update the *Guide* to include criteria for noise and vibration.
• Develop a performance standard for ventilated cages.
• Provide guidance to industry in a new facility design guide that can incorporate technological advances and current practices.

REFERENCES

AALAS [American Association of Laboratory Animal Science]. 2003. Evaluating odor migration in a new kennel project using CFD analysis. Poster presentation at the October 2003 meeting of the American Association for Laboratory Animal Science held in Seattle, Washington.
ASHRAE [American Society of Heating, Refrigerating and Air-Conditioning Engineers]. 2003. HVAC applications. In: ASHRAE Handbook. Atlanta, GA.

CDC [Centers for Disease Control and Prevention]. 1999. Biosafety in Microbiological and Biomedical Laboratories. Atlanta, GA: CDC.

NIH [National Institutes of Health]. 1999. NIH Design Policy and Guidelines. Bethesda, MD: NIH.

NRC [National Research Council]. 1996. Guide for the Care and Use of Laboratory Animals. 7th ed. Washington, D.C.: National Academy Press.

Riskowski, G.L., and F. Memarzadeh. 2000. Investigation of statis microisolators in wind tunnel tests and validation of CFD cage model. ASHRAE Trans 106:867-876.

Ruys, T., ed. 1991. Handbook of Facilities Planning. Vol 2. Laboratory Animal Facilities. New York: Van Nostrand Reinhold.

Sales, G., and D. Pye. 1974. Ultrasonic Communication by Animals. London: Chapman & Hall.

UFAW [Universities Federation for Animal Welfare]. 1996. Noise in Dog Kennnelling: A Survey of Noise Levels and the Causes of Noise in Animal Shelters, Training Establishments, and Research Institutions. Herts, UK: UFAW.

USDA [US Department of Agriculture]. 2002. ARS 242.1M. Washington, DC: Agricultural Research Service of USDA.

Warfield, D. 1973. The study of hearing in animals. In: Gay, W., ed. Methods of Animal Experimentation. London: Academic Press. p. 43-143.

European Guidelines for Environmental Control in Laboratory Animal Facilities

Harry J. M. Blom

Like farm animals and pets, laboratory animals were originally derived from wild living ancestors. The early scientists started to house and breed those animal species, mainly mammals, which were easiest to maintain under artificial conditions in terms of economics and animal needs. Of course, other criteria also played an important role in the selection process. The species of choice needed to be accurate models for biomedical research, the results of which were to be extrapolated to humans. Further easy breeding, a short life cycle, and large numbers of offspring were preferred—arguments that resulted in the use of rodents for experimental purposes. However, when introduced in the laboratory, the animals had to go through a process of habituation to the artificial housing conditions that far from resembled the animal's natural living environment. Animal enclosures in the modern animal facility are of a much better quality, and conditions are adequately controlled. Still, animals may be unable to adapt to these housing conditions and consequently may develop abnormal behavior, stress, affected physiology, and/or mental state. Therefore it is essential to define standards for housing conditions that meet the animals' requirements. Preferably these standards should be based on scientific data. Prevailing expert views and daily practice are to be considered acceptable when scientific data are not available.

The aim of international regulations for the care and use of laboratory animals is to enhance animal welfare, to set standards, to harmonize procedures, and to safeguard the quality of biomedical research. The European regulations for the protection of animals used for experimental and other scientific purposes are based on both scientific results and common sense. Directive 86/609/EEC provides mandatory guidelines for the 15 nations that are joined in the European Union. Convention ETS 123 has been set up by the 45 member states of the Council of Europe. The content of the Convention is mandatory in those member states that have signed and ratified this document. In both cases, the national authorities are obliged to transpose and implement the European regulations into national law. At this time, Appendix A to Convention ETS 123, providing guidelines for the housing and care of laboratory animals, as well as the European Directive are being revised. Other authors in these proceedings will elaborate on both revision processes. The focus of this presentation is on the new content of Appendix A.

After a special workshop in Berlin, Germany, in 1995, a Multilateral Consultation in 1997, and seven consecutive 3-day Working Party meetings in the period 1999-2003 at the Council of Europe in Strasbourg, France, discussion has been finalized for the General Part of Appendix A and for the species-specific sections for rodents and rabbits, dogs, cats, ferrets, nonhuman primates, and amphibians. It is anticipated that discussion on the sections for farm animals, birds, reptiles, and fish can be closed during the next meeting in early June 2004. Late in 2004, a Multilateral Consultation should conclude the revision process. The documents that are still under debate are restricted. The information presented herein is therefore limited to the finalized sections.

With respect to environmental conditions, the General Part contains provisions that are universally applicable to all laboratory animal species (Table 1). Where appropriate, the species-specific sections provide detailed guidelines, values, or ranges to meet the particular needs of the species concerned (Table 2). All provisions apply to inside enclosures. Where animals have access to outside enclosures, it is strongly recommended to prevent prolonged exposure to extreme climate conditions such as heat, frost, bright sunlight, or heavy rainfall. Although some species may tolerate such weather conditions relatively well, the animals should always have the ability to make a free choice to go inside or seek shelter.

As mentioned, the new sections in the revised Appendix A are based on scientific results. Unfortunately the availability of such data is limited. Thus, where science could not support the discussions during the Working Party meetings, there was no other option than to rely on expert views and common sense—a procedure that is fully justifiable but that emphasizes at the same time the need for further research into the tuning of housing conditions in the laboratory with the needs of the animals living in this artificial environment. The main problem to be solved is to generate

TABLE 1 Provisions on Environmental Conditions in the General Part of Appendix A to Council of Europe Convention ETS 123

Ventilation	• Should: — Satisfy the requirements of the animals — Provide sufficient fresh air of appropriate quality — Be 15-20 changes/hr — Remove excess heat and humidity — Prevent spread of odors, noxious gases, dust, and infectious agents • Recirculation of untreated air should be prevented • Draft and noise disturbance should be avoided
Temperature	• May affect metabolism and behavior of the animals • Should be precisely controlled (heat/cool) and measured and logged daily • Newborn, hairless, ill, and newly operated animals need special attention
Humidity	• May need to be controlled within a narrow range to minimize the possibility of health or welfare problems • Should be recorded and logged daily
Lighting	• Should satisfy biological requirements • Should provide a satisfactory working environment • Exposure to bright light should be avoided, and darker areas should be available • Regular photoperiods should be provided • Interruptions of the photoperiod should be avoided
Noise	• High noise levels and sudden loud noises may cause stress • Ultrasounds should be minimized particularly during the resting phase • Holding rooms should be provided with noise insulation and absorption materials

funding for these refinement studies. Furthermore, the classical approach of looking for signs of distress and/or discomfort evoked by imperfect housing conditions could be supplemented by studying expressions of pleasant experience. Predictability and controllability of the environment can be very rewarding to the animals and may therefore be expected to contribute to the well-being of captive living animals.

TABLE 2 Provisions on Environmental Conditions in the Species-Specific Sections of Appendix A to Council of Europe Convention ETS 123

	Temperature	Humidity	Lighting	Noise
Rodents	• 20-24°C (+6°C in cage) • Provide opportunity to control microclimate	• 55±10% • Gerbils 45±10%	• Low light levels in the cage • Albino's < 65 lux • Red light can be used for monitoring rodents in their active phase	• Are in particular very sensitive to ultrasound • Ultrasound may affect prenatal development • Sudden loud noises may cause audiogenic seizures
Rabbits	• 15-21°C (+6°C in cage) • Provide opportunity to control microclimate	• Not less than 45%		• Are in particular very sensitive to ultrasound • Ultrasound may affect prenatal development • Sudden loud noises may cause audiogenic seizures
Dogs	• 15-21°C when precise control is required during procedures • Otherwise a wider range provided that welfare is not compromised	• Control unnecessary • Can be exposed to wide fluctuations of ambient relative humidity without adverse effects	• Duration of the light period should be at least 10-12 hr • Low-level night lighting (5-10 lux) should be provided to avoid startle reflex	• Noise in dog kennels can reach high levels that can cause damage to humans and that could affect the dogs' health and physiology • By addressing the dogs' behavioral needs, barking may be decreased

continued

TABLE 2 Continued

	Temperature	Humidity	Lighting	Noise
Cats	• 15-21°C when precise control is required during procedures • Otherwise a wider range provided that welfare is not compromised	• Control unnecessary • Can be exposed to wide fluctuations of ambient relative humidity without adverse effects	• Duration of the light period should be at least 10-12 hr • Low-level night lighting (5-10 lux) should be provided to avoid startle reflex • Light sources may be perceived as flickering because of the cats' high *critical fusion frequency*	• Unpredictable noise may cause stress
Ferrets	• 15-24°C • Absence of well-developed sweat glands may lead to risk of heat exhaustion when exposed to high temperatures	• Control unnecessary • Can be exposed to wide fluctuations of ambient relative humidity without adverse effects	• Duration of the light period may vary between 8-16 hr • Modification of the photoperiod is an important tool for the manipulation of the reproductive cycle	• Lack of sound or auditory stimulation can be detrimental and make ferrets nervous • Loud unfamiliar noise and vibration have been reported to cause stress-related disorders

continued

TABLE 2 Continued

	Temperature	Humidity	Lighting	Noise
Nonhuman Primates				
• **Marmosets and Tamarins**	• 23-28°C but slightly higher levels are acceptable	• 40-70% but levels higher than 70% will be tolerated	• Not less than 12 hr of light • Provision of a shaded area	• Exposure to ultrasound should be minimized
• **Squirrel Monkeys**	• 23-28°C without abrupt temperature variations	• 40-70%	• Not less than 8 hr of light • Light spectrum should resemble daylight, i.e. including UV light	
• **Macaques and Vervets**	• 16-25°C is suitable • 21-28°C is more suitable for long-tailed macaques	• 40-70%	• 12:12 light/dark cycle	
• **Baboons**	• 16-28°C is suitable			

continued

TABLE 2 Continued

	Temperature	Humidity	Lighting	Noise
Amphibians	• Amphibians are ectothermic • Areas of different temperatures are beneficial • Exposure to frequent fluctuations in temperature should be avoided	• A hydrated integument and the possibility to take up moisture through the skin are essential	• Photo-periods and light intensities should be consistent with the natural conditions	• Noise, vibration, and unexpected stimuli should be minimized
• **Aquatic urodeles**	• 15-22°C	• 100%		
• **Aquatic anurans**	• 18-20°C	• 100%		
• **Semi-aquatic anurans**	• 8-10°C	• 50-80%		
• **Semi-terrestrial anurans**	• 23-27°C	• 80%		
• **Arboreal anurans**	• 18-25°C	• 50-70%		

NOTE: For ventilation the provisions in the General Part apply to all species. The same applies for empty cells in the table.

Breakout Session: Lighting

Leader: Harry J. M. Blom

Rapporteur: Michael K. Stoskopf

The session leader posed the following questions to the group:

What is the scientific basis or peer-reviewed literature for housing standards for laboratory animals? What other (if any) influences or factors are involved?

General consensus was quickly reached among the participants that the scientific basis for housing standards for laboratory animals is uneven, with large areas that lack adequate investigation. Some concern was expressed that not all "scientific" information is sufficient for development of standards. Participants stressed the care in design and execution of experiments, including the need for replication and proper controls necessary to provide reliable information. In addition, scientific design and replication of studies varies: One poorly designed study can dictate standards inappropriately.

The moderator posited that the influences on standards, other than peer-reviewed scientific data, include daily practice, common sense, and prevailing expert views. It was suggested that it might be appropriate to establish standards. The group allowed that although these factors do become the basis of standards, there are important concerns with this approach, and the development of standards without scientific basis is fraught with the peril of inappropriate regulation. These concerns were expressed first in a question related to expertise and second in the subsequent brief discussion of common sense.

Who determines the prevailing expert view?

Participants indicated that individuals with different backgrounds can have diametrically opposed biases on appropriate prioritization of the various concerns. In addition, they felt that giving unfounded dogma official sanction can retard proper scientific examination of the viewpoint. Similarly, some expressed the opinion that common sense can frequently be wrong. Individuals may tend to express anthropomorphic views basing judgment of other species, needs on human senses and needs, and this may not always be appropriate.

Although it is necessary to have some basis for a starting point, participants felt it would be optimal if the starting point were based on scientific evidence.

Where are the gaps in our scientific knowledge? Is the information missing? Is it outdated?

Discussion of these questions was divided into three areas related to light: (1) intensity, (2) periodicity, and (3) transitions, including "flicker detection."

INTENSITY

Issues related to light intensity were organized into the following three categories for discussion: (1) satisfaction of biological requirements, (2) safety and efficiency of people working in the room, and (3) effects of excessive exposure. All three areas have gaps in knowledge.

Biological requirements related to vitamin synthesis have been determined for some species; however, less is known about intensity requirements relative to neuroendocrine function, especially across a broadly comparative group of laboratory species. The safety and efficiency of people working in the room have been studied more than the preceding category, but often in studies unrelated to laboratory animal care. Human effectiveness and its variability under different lighting conditions are relatively well studied. Much of the discussion focused on the issue of effects of exposure to excessive intensity, with particular focus on light that is too bright and causes blindness or retinal lesions in some species. Very bright light should be avoided for some species (e.g., albino rodents, as recommended in the *Guide* reference to 30 to 50 foot-candles), and darker areas should be available to the animals. With regard to the needs of other strains and species, participants stated that tiered cages and the location in the tier are factors that have not been considered, because most studies have looked at average room conditions.

Some time during this session was devoted to determining how different institutions are dealing with light intensity challenges. Mentioned were manufactured lenses covering fluorescent bulbs to reduce lighting to within range; removal of some of the bulbs commonly used for lights in ceiling, but not with uncovered tops on racks; and the practice of rearing animals in the dark (e.g., use of transgenics) in ophthalmology studies. Participants expressed concerns regarding potential damage caused by light intensity that is too low (e.g., on the retina).

PERIODICITY

Dr. Blom suggested starting with the following areas, in which the knowledge base is well established:

- Providing regular photoperiods;
- Avoiding interruptions of those periods;
- Considering low-level night lighting; and the
- Potential importance of the duration of the light/dark cycle, for the manipulation of reproductive cycles in breeding and related research.

The group did not take exception to those points, but discussed that periodicity and particularly the duration of light/dark cycles is important for many other things besides reproduction.

Much of the discussion centered on experiences with nocturnal animals such as owl monkeys. There is still more to be understood about the use of simulated moonlight (lower intensity vs. spectral shifts) across species. Simulated moon light is being practiced in some of the forms, but has proven impractical for allowing workers to properly clean and manage rooms and adaptation to the low levels (about 10 lux), For this reason, it does not appear to be effective. The main solution to the problem created by workers being required to turn on the lights appears to be creative shifts of time reversal so that "daylight" exposures occur during working and cleaning, and "dark" periods are reserved for observation periods.

The important point was stressed that this issue is more refined than simply identifying the light/dark cycle. The cycle can affect results for many types of studies such as metabolism, for which considerable data exist. In addition, it has been shown that seasonal shifts in diurnal cycles can be crucial.

The group identified an important need for better reporting of husbandry and procedures in published papers to allow evaluation and replication of the studies. It is hoped that online publishing will help resolve this problem, but participants recognize that a strong demand for complete disclosure of husbandry, including light management, is needed.

TRANSITIONS

Our knowledge of the impact of transitions is weak, but we have reasonable scientific basis from field studies to consider that they may affect research outcomes and perhaps animal wellness. Rapid transitions will invoke alarm behaviors in several species, and the metabolic impacts of these transitions are poorly understood. Although considerable effort has been invested in managing diurnal cycles in some species and facilities, much less effort has gone into managing transitions. One possible challenge has been the wide spread use of fluorescent lighting, which requires relatively expensive electronics to dim. Those devices have also recently been shown to produce ultrasound at levels that could be problematic. The consensus of participants is that expense was the main driving force in the use of fluorescent lighting, and this in turn has resulted in limited options in light management.

WAVELENGTH/FREQUENCY

Dr. Blom posed the possibility that rodents could use red light during their active phase, which would also constitute a good approach for balancing the need for humans to see during the active phase. In this context, participants indicated the existence of gaps in the following areas:

- The effect of red light;
- Whether blue light is more appropriate for nocturnal periods;
- The need to identify the ultraviolet (UV) requirements of various species (already known for some reptiles, birds, and insects, but largely extrapolated across mammals); and
- Whether animals need exposure to a spectrum of full daylight.

FLICKERING

The issue of flickering was discussed because of challenges identified in Europe. Because 50 Hz is used as the typical cycle for power in Europe, fluorescent bulbs flicker at 50 Hz, rather than at 60 Hz, the cycle commonly used in the United States. The critical fusion frequency for an individual or species is the frequency at which a cycling light would be perceived as a continuous light source. The higher the critical fusion frequency of a species, the more likely they would be to perceive a fluorescent light as flickering on and off rather than providing steady light. This problem also occurs in humans and is the basis of considerable investigation relative to impacts on health and well-being. For cats, the critical fusion frequency is known to be slightly >50 Hz. For birds, the problem may be more acute because birds have a fourth type of cone that is sensitive to UV light and is phasic (i.e., sensitive to flickering up to ≥110 Hz).

Some participants suggested that potential problems from flicker perception should be studied in laboratory species. Others identified additional problems from fluorescent lights including generation of ultrasound by ballasts.

GAPS IN THE KNOWLEDGE

Among the many knowledge gaps related to light and laboratory species, the following areas clearly require additional information:

- Knowledge of species and strain variations in susceptibilities and needs;
- Natural history studies with communication to laboratory animal scientists (e.g., metabolic shifts, behavioral endocrine shifts);
- Photoperiodicity studies;
- Light acuity sensitivity data;
- Studies on the effects of maintaining rodents under dim light with periods of increased light intensity; and
- Studies of the effects of cage materials (e.g., clear vs. tinted walls) at actual light levels experienced in the cage itself (secondary enclosures) as opposed to the room.

ENGINEERING STANDARDS

Some participants felt that it is possible to spend so much time and effort on a particular engineering standard that time working with animal enrichment is severely decreased. Participants identified prioritization of effort as an important issue.

Similarly, it was felt that engineering standards can create important problems if based on poor data. This problem occurs particularly when engineering standards are too tightly defined and result in retarding the generation of new knowledge. It is common for engineering standards to conflict with performance standards.

EFFECT OF CURRENT REGULATIONS ON THE WELFARE OF THE ANIMALS

"Shoulds" tend to evolve into "musts." There is a strong need for the use of adaptive management in many laboratory animal maintenance situations. In the absence of knowledge, the freedom to experiment and explore options is required. Participants indicated that investigators should be encouraged to study the effects of husbandry on their research.

It is possible for IACUC chairs to prefer strong and narrow regulations with tight interpretations to facilitate their ability to exert control over investigators who are not in optimal compliance. That approach, of functioning as a policeman and an enforcer, is one alternative; however, the approach of working with investigators as part of a team to improve animal welfare and care seems to be more effective.

IDENTIFICATION OF SIGNIFICANT DIFFERENCES AND CONFLICT (IF ANY) IN GUIDELINES/STANDARDS

This issue was not addressed during the session in detail because of time constraints, but the general position of the participants was to embrace performance-based standards in preference to specific engineering standards. This position was based largely on the perceived need to address a wide range of species and strains that may have different needs. Also of concern was the need to balance the lighting needs of animals with those of staff who are attempting to maintain the colony or conduct research.

Breakout Session:
Effects of Sound on Research Animals

Leader: Sherri L. Motzel

Rapporteur: Hilton J. Klein

Gaps in our knowledge exist regarding the effects of noise, vibration, and sound for research animals. In this session, Dr. Sherri Motzel, Director of Laboratory Animal Resources at Merck Research Laboratories, presented a scholarly review of the effects of noise, vibration, and sound. The review included definitions, the current regulations and standards for noise, reviews of several relevant studies for rodents and nonhuman primates, and opportunities for noise, sound, and vibration mitigation. Dr. Motzel provided several references and cited relevant studies demonstrating that noise, vibration, and sound can have deleterious effects on behavioral and physiological parameters (Motzel et al. 2001; Sales and Milligan 1992; Sales et al. 1998, 1999).

Sound, which is produced when vibrating objects cause changes in air pressure, varies in duration, frequency (Hz), and magnitude or intensity (decibels, sound pressure level). Laboratory animals vary greatly by species in their ability to detect sound compared with humans. For example, humans detect sound from 20 Hz to 20 kHz, whereas rodent species are much more diverse in their ability to detect sound. Examples of range of detection include the following:

Mouse	0.8–100 kHz
Rat	0.25-76 kHz
Nonhuman primate (rhesus)	0.13-45 kHz
Dog	0.04-46 kHz

Thus, animals detect sound inaudible to humans.

The US Animal Welfare Act regulations do not address noise. However, the ILAR *Guide for the Care and Use of Laboratory Animals* (NRC 1996) includes the recommendation to assess the effects of noise on animals, and to consider noise controls in animal facility design and construction. The Agricultural *Guide* (APHIS 1998) reflects greater tolerance toward the effects of noise on farm/agricultural animals, based on the few permanent effects reported in the literature cited therein. The Agricultural *Guide* does, however, include the recommendation that noise control should be considered during facility design. In contrast, the Council of Europe is clearer about the stressful effects of noise on laboratory animals and provides more specific recommendations in noise mitigation and control as well as for facility layout, design, and construction. In summary, regulations and standards for all laboratory animals address noise in a very basic and fundamental manner, yet they do not address the noise issue extensively because of a paucity of data on noise effects in the peer-reviewed literature.

Dr. Motzel reviewed sources and types of sound and their effects in animal laboratory settings. Recorded sound levels vary widely but are dependent on species (e.g., barking dogs—99 dB) and on work practices, work cycles, and equipment.

Ultrasonic sound has been recorded from 24 of 39 sources (e.g., video displays, furniture, vacuums, and cage washers) and in some cases exceeds 100 kHz and 122 dB in frequency and intensity. It has been demonstrated clearly that ultrasonic sound creates perturbations in physiological parameters (e.g., heart rate, blood pressure, electroencephalographic changes), behavioral parameters (seizing), and teratogenic effects on laboratory animals.

Sound effects also vary in their impact, depending on the animal species, strain, and age. Dr. Motzel cited clear-cut effects of sound on response to drug treatment, water intake, blood pressure, reproduction, glucose metabolism, and immune function. One study conducted at Merck Research Laboratories by Dr. Motzel and her colleagues demonstrated conclusively that infrasound (1-10 Hz) was responsible for weight loss in CD rats in the study. A malfunctioning air handler was responsible for the source of the subsonic noise, which caused the weight loss. This study and other reports in the literature indicate that much more emphasis should be placed on monitoring and controlling noise levels at multiple frequency and intensity ranges outside human hearing ranges in animal facilities because of the potential for adverse effects on study data and outcomes. Preventive maintenance and facility testing, facility design, and work practices should also be reassessed in the laboratory animal facility in an effort to mitigate adverse noise effects. It was suggested that these strategies are effective for control and mitigation. Sound neutralizers and sound breaks were briefly mentioned as control devices for excessive noise problems.

In summary, the group agreed with Dr. Motzel's assessments that in the context of behavioral and physiological effects, some laboratory animals are more sensitive to noise than humans. These effects are observed across a range of frequency, intensity, and duration that is much broader than in humans. Participants believed that for this reason, the current standards for the human environment may be of limited relevancy and not adequate to protect the integrity of research experiments. Additional in-depth review of the literature combined with relevant research studies to address noise effects in laboratory animals is clearly indicated. Participants agreed that current regulations and guidelines should be revised and updated accordingly.

SELECTED REFERENCES

APHIS [Animal Plant Health Inspection Service]. 1998. Regulation of agricultural animals (policy 26). In: Animal Care Resource Guide. Washington, DC: US Department of Agriculture.

Motzel, S.L., Morrisey, R.E., Conboy, T.A., and others. 1996. Weight loss in rats associated with exposure to infrasound. Contemp Topics 35:69.

NRC [National Research Council]. 1996. Guide for the Care and Use of Laboratory Animals. 7th ed. Washington, DC: National Academy Press.

Sales, G.D., and S.R. Milligan. 1992. Ultrasound and laboratory animals. Anim Technol 43:89-98.

Sales, G.D., Milligan, S.R., Khirnykh, K. 1999. Sources of sound in the laboratory animal environment: A survey of the sounds produced by procedures and equipment. Anim Welfare 8:97-115.

Sales, G.D., Wilson, K.J., Spencer, K.E.V., Milligan, S.R. 1998. Environmental ultrasound in laboratories and animal houses: A possible cause for concern in the welfare and use of laboratory animals. Lab Anim 22:369-375.

Breakout Session: Environmental Control for Animal Housing— Impact on Metabolism and Immunology

Leaders: Jann Hau and Randall J. Nelson

Rapporteur: Stephen W. Barthold

The introductory discussion focused on the impact of new guidelines on immune response and metabolism. Significant changes that may influence these responses include social grouping, environmental enrichment, and enclosure size.

Questions:

1. Is there consistent scientific evidence for an impact of social environment (or environmental enrichment) on the immune system and metabolism?

If so, is the evidence species specific?

2. Is there a need for additional research on the impact of social environment or environmental enrichment on immune system and metabolism?

If yes, in which areas?

3. Is it possible to produce guidelines (or "best practices") for group sizes of different species (strains, sexes, age groups) that would be optimal (i.e., not cause added variation to immune system and metabolism parameters)?

Does the answer to this question depend on the project or parameter studied?

4. Will there be a need for single housing to control variation with respect to immune system and metabolism?

5. To develop "best practices" how should the group categorize the species (e.g., rodents vs. primates; solitary vs. social)?

6. Is it necessary to code these "best practices," or is a mixed model of voluntary AND regulatory practices most sensible?

7. Is it necessary to distinguish clearly between "stock and use" situations (i.e., one type of housing for the "hotel" period and another type of housing for the "experiment" period)?

8. Is it necessary to consider the case of breeding colonies?

Drs. Hau and Nelson presented the following specific situations involving primates that illustrate these issues:

- Single-housed gorillas have elevated cortisol (Stoinsky and others 2002).
- Single housing of rhesus causes long-term immunosuppression (Lilly and others 1999).
- Single housing of African green monkeys induces immuno-suppression (Suleman and others 1999).
- Pair housing of marmosets reduces cortisol response to novelty (Smith and others 1998).
- Social separation of cynomolgus monkeys exacerbates athero-sclerosis (Watson and others 1998).
- Transfer from natal group to peer group of juvenile rhesus affects cortisol and T cell subsets (Gust and others 1992).
- Separation of juvenile rhesus from natal group induces immuno-suppression (Gordon and others 1992).
- Formation of unrelated rhesus females into groups induces immuno-suppression (Gust and others 1991).
- Social group stress induces endothelial dysfunction in cynomolgus monkeys (Strawn and others 1991).

Examples of situations involving rats include the following:

- Isolation advances puberty; enrichment delays puberty (Swanson and van de Poll 1983).
- Single housing impairs testosterone synthesis and produces Leydig cell atrophy (Nyska and others 1998); increased exercise induces weight loss (Boakes and Dwyer 1997).
- Group-housed rats are less stressed than single-housed (Sharp and others 2002, 2003) but are more vulnerable to stress-induced ulcers (Pare and others 1985).
- Individually reared rats have a less than adequate response to aggression (Von Frijtag and others 2002).
- Single housing (accompanied by stress) does not reduce the immune response of the rat to an antigen (Baldwin and others 1995).

- Single-housed rats are characterized by:
  - Higher levels of cortisol and prolactin (Gambardella and others 1994);
  - Increased substance P in the spinal cord (Brodin and others 1994);
  - Reduction of hypertension in obese rats; and
  - Reduced tumor growth (Steplewski and others 1987).

Examples of situations involving laboratory rodents include the following:

- Single housing does not change glucocorticoid concentrations (Benton and Brain 1981; Misslin and others 1982) and does not affect reaction to a stressor (immunosuppression; Bartolomucci and others 2003).
- Single housing induces immunosuppression (Shanks and others 1994).
- Crowding males potentiates corticosterone response to acute stress (Laviola and others 2002).
- Single-housed rats and mice behave differently in behavioral tests (Karolewicz and Paul 2001; Palanza and others 2001).
- Minimal stress with four mice per cage compared with two or eight per cage (Peng and others 1989).
- Male aggression is greater in groups of eight than in groups of three to five. Decreasing floor space decreases aggression (Van Loo and others 2001).

Social housing influences have included the following:

- Expression of heat shock proteins (Andrews and others 2000);
- Corticotropin-releasing factor and GABA receptors (Matsumoto and others 1997);
- Chemotherapeutic efficacy (Kerr and others 1997, 2001);
- Tumor growth (Kerr and others 2001; Rowse and others 1995; Weinberg and Emerman 1989);
- Streptozotocin-induced hyperglycemia (Mazelis and others 1987); and
- Hematopoiesis (Williams and others 1986).

Experience from immunization has been documented. Group-housed males have:

- Higher cortisol levels and are immunosuppressed compared with single housed and family housed;

- Primary response to an antigen is low but response to booster injection is normal (Abraham and others 1994);
- No reports on difference in titer (e.g., development in rabbits housed in different social groups).

Examples of the effect of enrichment include the following:

- Barren-housed pigs have impaired long-term memory and blunted circadian cortisol rhythm (de Jong and others 2000).
- Environmental enrichment stimulates the hypothalamic-pituitary-adrenal axis and the immune system in mice (Marashi and others 2003).
- Numerous examples of the positive effect of enrichment on brain function are in the literature (e.g., reviews by Larsson and others 2002; Mattson and others 2001; Risedal and others 2002; Schrijver and others 2002).

VARIANCE IN EXPERIMENTAL RESULTS

The contribution of enrichment to variance in experimental results appears to depend on respective parameters. Dr. Nelson discussed the issue of density of animals in a room, using species variation as an example. He discussed data that indicated high-density populations result in high steroid concentrations and decreased immune function in mice (e.g., Csermely and others 1995; Tsukamoto and others 1994), but increased immune function in prairie voles (Nelson and others 1996). Thus, he concluded that intuition cannot be used in establishing guidelines, which reinforces the concept that guidelines must be science based and species sensitive.

Dr. Nevalainen presented data regarding volatile compounds in bedding, and environmental enrichment with variable material. Some bedding materials contain chemicals known as pinenes, which are heat labile, but induce hepatic microsomal enzymes (Nevalainen and Vartainen 1996). He also emphasized the need to utilize consistent materials for enrichment that are inert to other environmental materials to which the animals are exposed.

DISCUSSION AND POINTS RAISED BY PARTICIPANTS

One participant raised the issue of diet as another environmental variable that is not well controlled. There is a trend to replace some ingredients with others, such as replacing fish protein with casein. It was also noted that there is a growing number of rodents with suppressed immune responses due to highly hygienic husbandry practices among commercial breeders. Dr. Hardy cited an instance in which BALB/cByJ mice have a 4- to 10-fold decrease in total immunoglobulin (Ig)G, decreased mass to organized lymphoid tissue, and resulting shifts in immune reactivity. In highly sensitive studies, there is an increased trend of using gnotobiotic mice that are significantly affected by this phenomenon of immune system hypoplasia. For this reason, it was felt that it is important to determine the standards for rodent microflora. Users have a high sensitivity to issues relating to "microbial drift" in breeder colonies, which in turn have a large impact on breeders. Some strains of mice, including many transgenic mice, are more sensitive than others to these effects. Other effects of this immune hypoplasia syndrome include plasmacytogenesis in BALB mice primed with pristane, in which the mice have decreased yield and primarily IgM, rather than IgG. Susceptibility to other infections such as *Giardia* is also seen.

Biological endpoints have changed drastically and are generally more sensitive. Thus, the impact of environmental variables becomes more obvious and poses challenges for high-throughput analyses. How long do animals need to acclimate before being placed in test environments? Most people use a range of 24 hours to 5 days, but there have been no new data for more than 20 years. The animals may never acclimatize, such as when they are singly housed after having been maintained in a group.

Many participants indicated that guidelines and regulations are not the answer. The Materials and Methods sections in scientific publications must provide documentation of the study design, including such variables. Unfortunately, journals encourage less, rather than more, detail, which reduces the reproducibility of science and increases unnecessary use of animals to obtain reproducible results in other laboratories. The underlying principle is that "variance varies with various variables," and guidelines or regulations with straight and narrow standards or limits interfere with this concept.

When discussing the possibility of developing guidelines for optimal group sizes, participants indicated that consideration needs to be given to factors such as species, sex, and strain, which make such rigid guidelines impossible. The answer depends on the project, and science must guide science, not rigid regulations. The current *Guide* (NRC 1996) dictates the number of mice per unit area of cage, and these guidelines, which are not based on science, are still often used as rigid standards. Considerable discussion revolved around the fact that the *Guide* is a guide, and that it is being misused by regulators. More details in any new iterations of the *Guide* will likely create more rules, without real benefit to animals or science. Dr. White's presentation accurately depicted the reality. The *Guide* has only three musts in the entire book. IACUCs and regulatory agencies need better education regarding the purpose and limitations of the *Guide*.

The Canadians seem to be doing the best, with a highly flexible and adaptive system of guidelines and oversight.

Finally, discussion of enriched versus nonenriched environments continued. Moving animals from unenriched production environments to different enriched environments for holding, then nonenriched environments for experimentation, creates enormous variation in response. Thus, it was felt that consideration must be given to the impact of new guidelines that may be well intentioned but not based on science and their impact on science.

REFERENCES

Abraham, L., O'Brien, D., Poulsen, O.M., Hau, J. 1994. The effect of social environment on the production of specific immunoglobulins against an immunogen (human IgG) in mice. In: Bunyan, J., ed. Welfare and Science. London: Royal Society of Medicine Press. p. 165-170.

Andrews, H.N., Kerr, L.R., Strange, K.S., Emerman, J.T., Weinberg, J. 2000. Effect of social housing condition on heat shock protein (HSP) expression in the Shionogi mouse mammary carcinoma (SC115). Breast Cancer Res Treat 59:199-209.

Baldwin, D.R., Wilcox, Z.C., Baylosis, R.C. 1995. Impact of differential housing on humoral immunity following exposure to an acute stressor in rats. Physiol Behav 57:649-653.

Bartolomucci, A., Sacerdote, P., Panerai, A.E., Peterzani, T., Palanza, P., Parmigiani S. 2003. Chronic psychosocial stress-induced down-regulation of immunity depends upon individual factors. J Neuroimmunol 141:58-64.

Benton, D., and Brain, P.F. 1981. Behavioral and adrenocortical reactivity in female mice following individual or group housing. Dev Psychobiol 14:101-107.

Boakes, R.A., and Dwyer, D.M. 1997. Weight loss in rats produced by running: effects of prior experience and individual housing. Q J Exp Psychol [B] 50:129-148.

Brodin, E., Rosen, A., Schott, E., Brodin, K. 1994. Effects of sequential removal of rats from a group cage, and of individual housing of rats, on substance P, cholecystokinin and somatostatin levels in the periaqueductal grey and limbic regions. Neuropeptides 26:253-260.

Csermely, P., Penzes, I., Toth, S. 1995. Chronic overcrowding decreases cytoplasmic free calcium levels in T lymphocytes of aged CBA/CA mice. Experientia 51:976-979.

de Jong, I.C., Prelle, I.T., van de Burgwal, J.A., Lambooij, E., Korte, S.M., Blokhuis, H.J., Koolhaas, J.M. 2000. Effects of environmental enrichment on behavioral responses to novelty, learning, and memory, and the circadian rhythm in cortisol in growing pigs. Physiol Behav 68:571-578.

Gambardella, P., Greco, A.M., Sticchi, R., Bellotti, R., Di Renzo, G. 1994. Individual housing modulates daily rhythms of hypothalamic catecholaminergic system and circulating hormones in adult male rats. Chronobiologia Int 11:213-221.

Gordon, T.P., Gust, D.A., Wilson, M.E., Ahmed-Ansari, A., Brodie, A.R., McClure, H.M. 1992. Social separation and reunion affects immune system in juvenile rhesus monkeys. Physiol Behav 51:467-472.

Gust, D.A., Gordon, T.P., Wilson, M.E., Brodie, A.R., Ahmed-Ansari, A., McClure, H.M. 1992. Removal from natal social group to peer housing affects cortisol levels and absolute numbers of T cell subsets in juvenile rhesus monkeys. Brain Behav Immun 6:189-199.

Karolewicz, B., and Paul, I.A. 2001. Group housing of mice increases immobility and anti-depressant sensitivity in the forced swim and tail suspension tests. Eur J Pharmacol 415:197-201.

Kerr, L.R., Grimm, M.S., Silva, W.A., Weinberg, J., Emerman, J.T. 1997. Effects of social housing condition on the response of the Shionogi mouse mammary carcinoma (SC115) to chemotherapy. Cancer Res 57:1124-1128.

Kerr, L.R., Hundal, R., Silva, W.A., Emerman, J.T., Weinberg, J. 2001. Effects of social housing condition on chemotherapeutic efficacy in a Shionogi carcinoma (SC115) mouse tumor model: influences of temporal factors, tumor size, and tumor growth rate. Psychosomat Med 63:973-984.

Larsson, F., Winblad, B., Mohammed, A.H. 2002. Psychological stress and environmental adaptation in enriched vs. impoverished housed rats. Pharmacol Biochem Behav 73:193-207.

Laviola, G., Adriani, W., Morley-Fletcher, S., Terranova, M.L. 2002. Peculiar response of adolescent mice to acute and chronic stress and to amphetamine: evidence of sex differences. Behav Brain Res 130:117-125.

Lilly, A.A., Mehlman P.T., Higley, J.D. 1999. Trait-like immunological and hematological measures in female rhesus across varied environmental conditions. Am J Primatol 56:73-87.

Marashi, V., Barnekow, A., Ossendorf, E., Sachser, N. 2003. Effects of different forms of environmental enrichment on behavioral, endocrinological, and immunological parameters in male mice. Horm Behav 43:281-292.

Matsumoto, K., Ojima, K., Watanabe, H. 1997. Central corticotropin-releasing factor and benzodiazepine receptor systems are involved in the social isolation stress-induced decrease in ethanol sleep in mice. Brain Res 753:318-321.

Mattson, M.P., Duan, W., Lee, J., Guo, Z. 2001. Suppression of brain aging and neuro-degenerative disorders by dietary restriction and environmental enrichment: molecular mechanisms. Mech Ageing Dev 122:757-778.

Mazelis, A.G., Albert, D., Crisa, C., Fiore, H., Parasaram, D., Franklin, B., Ginsberg-Fellner, F., McEvoy, R.C. 1987. Relationship of stressful housing conditions to the onset of diabetes mellitus induced by multiple, sub-diabetogenic doses of streptozotocin in mice. Diabetes Res 6:195-200.

Misslin, R., Herzog, F., Koch, B., Ropartz, P. 1982. Effects of isolation, handling and novelty on the pituitary—adrenal response in the mouse. Psychoneuroendocrinology 7:217-221.

Nelson, R.J., Fine, J.M., Moffatt, C.A., Demas, G.E. 1996. Photoperiod and population density interact to affect reproductive, adrenal, and immune function in male prairie voles (*Microtus ochrogaster*). Am J Physiol 270:R571-R577.

Nevalainen, T., and Vartainen, T. 1996. Volatile organics compounds in commonly used beddings before and after autoclaving. Scand J Lab Anim Sci 23:101-104.

NRC [National Research Council]. 1996. Guide for the Care and Use of Laboratory Animals. 7th ed. Washington, DC: National Academy Press.

Palanza, P., Parmigiani, S., vom Saal, F.S. 2001. Effects of prenatal exposure to low doses of diethylstilbestrol, o,p'DDT, and methoxychlor on postnatal growth and neuro-behavioral development in male and female mice. Horm Behav 40:252-265.

Pare, W.P., Vincent, G.P., Natelson, B.H. 1985. Daily feeding schedule and housing on inci-dence of activity-stress ulcer. Physiol Behav 34:423-429.

Peng, X., Lang, C.M., Drozdowicz, C.K., Ohlsson-Wilhelm, B.M. 1989. Effect of cage popu-lation density on plasma corticosterone and peripheral lymphocyte populations of laboratory mice. Lab Anim 23:302-306.

Risedal, A., Mattsson, B., Dahlqvist, P., Nordborg, C., Olsson, T., Johansson, B.B. 2002. Environmental influences on functional outcome after a cortical infarct in the rat. Brain Res Bull 58:315-321.

Rowse, G.J., Weinberg, J., Emerman, J.T. 1995. Role of natural killer cells in psychosocial stressor-induced changes in mouse mammary tumor growth. Cancer Res 55:617-622.

Schrijver, N.C., Bahr, N.I., Weiss, I.C., Wurbel, H. 2002. Dissociable effects of isolation rearing and environmental enrichment on exploration, spatial learning and HPA activity in adult rats. Pharmacol Biochem Behav 73:209-224.

Shanks, N., Renton, C., Zalcman, S., Anisman, H.1994. Influence of change from grouped to individual housing on a T-cell-dependent immune response in mice: antagonism by diazepam. Pharmacol Biochem Behav 47:497-502.

Sharp, J.L., Zammit, T.G., Azar, T.A., Lawson, D.M. 2002. Stress-like responses to common procedures in male rats housed alone or with other rats. Contemp Topics Lab Anim Sci 41:8-14.

Sharp, J., Zammit, T., Azar, T., Lawson, D. 2003. Stress-like responses to common procedures in individually and group-housed female rats. Contemp Topics Lab Anim Sci 42:9-18.

Smith, T.E., McGreer-Whitworth, B., French, J.A. 1998. Close proximity of the heterosexual partner reduces the physiological and behavioral consequences of novel-cage housing in black tufted-ear marmosets (Callithrix kuhli). Horm Behav 34:211-222.

Steplewski, Z., Goldman, P.R., Vogel, W.H. 1987. Effect of housing stress on the formation and development of tumors in rats. Cancer Lett 34:257-261.

Stoinski, T.S., Czekala, N., Lukas, K.E., Maple, T.L. 2002. Urinary androgen and corticoid levels in captive, male Western lowland gorillas (Gorilla g. gorilla): Age- and social group-related differences. Am J Primatol 56:73-87.

Strawn, W.B., Bondjers, G., Kaplan, J.R., Manuck, S.B., Schwenke, D.C., Hansson, G.K., Shively, C.A., Clarkson, T.B. 1991. Endothelial dysfunction in response to psychosocial stress in monkeys. Circ Res 68:1270-1279.

Suleman, M.D., Yole, D., Wango, E., Sapolsky, R., Kithome, K., Carlsson, H.E., Hau, J. 1999. Peripheral blood lymphocyte immunocompetence in wild African green monkeys (Cercopithecus aethiops) and the effects of capture and confinement. In Vivo 13:25-27

Swanson, H.H., and van de Poll, N.E. 1983. Effects of an isolated or enriched environment after handling on sexual maturation and behaviour in male and female rats. J Reprod Fertil 69:165-171.

Tsukamoto, K., Machida, K., Ina, Y., Kuriyama, T., Suzuki, K., Murayama, R., Saiki, C. 1994. Effects of crowding on immune functions in mice. Nippon Eiseigaku Zasshi 49:827-836.

Van Loo, P.L., Mol, J.A., Koolhaas, J.M., Van Zutphen, B.F., Baumans, V. 2001. Modulation of aggression in male mice: influence of group size and cage size. Physiol Behav 72:675-683.

Von Frijtag, J.C., Schot, M., van den Bos, R., Spruijt, B.M. 2002. Individual housing during the play period results in changed responses to and consequences of a psychosocial stress situation in rats. Dev Psychobiol 41:58-69.

Watson, S.L., Shively, C.A., Kaplan, J.R., Line, S.W. 1998. Effects of chronic social separation on cardiovascular disease risk factors in female cynomolgus monkeys. Atherosclerosis 137:259-266.

Weinberg, J., and Emerman, J.T. 1989. Effects of psychosocial stressors on mouse mammary tumor growth. Brain Behav Immun 3:234-246.

Williams, L.H., Udupa, K.B., Lipshitz, D.A. 1986. Evaluation of the effect of age on hematopoiesis in the C57BL/6 mouse. Exp Hematol 14:827-832.

Breakout Session:
Environmental Control/
Engineering Issues

Leaders: Bernard Blazewicz and Dan Frazier

Rapporteur: Janet Gonder

Participants began by listing a series of issues for possible discussion. Topics included biosafety and biosecurity; ventilation rates and effectiveness of ventilation; ventilated caging systems; relative humidity control; sources of humidification; monitoring; need for filtration; and sources of contaminants. Several of these issues were discussed.

Discussion of the engineering issues related to biohazard research centered around biosafety level (BSL) 3 and BSL4 housing for agricultural animals and nonhuman primates. The impetus for this discussion is the new funding for facility construction for national and regional containment laboratories and other research programs. Current design and construction references include the Centers for Disease Control and Prevention (CDC/NIH) Biosafety in Microbiological and Biomedical Laboratories (BMBL), National Institutes of Health (NIH) Guidelines and Policies, and the US Department of Agriculture (USDA) document ARS 242.1M. Of the many design considerations for this type of facility, it has become apparent that interpretation of the requirements is changing as experience is gained with these facilities. Participants identified a need for tracking these changes and experiences. It was noted that some of this information is available through the American Biological Safety Association (ABSA). The need for consideration of system redundancy was discussed, along with the need for more information to enable institutions to perform adequate risk assessments to maintain safety. It was also pointed out that there is a critical need for knowledgeable engineering staff in these facilities to monitor and maintain the systems.

Ventilation rates of 10 to 15 air changes per hour (ac/h) have been cited in the *Guide for the Care and Use of Laboratory Animals* (*Guide*) as a reasonable general guidance. However, it was agreed that in some circumstances less than 10 ac/h may be adequate, whereas in other cases, more than 15 ac/h are needed to address the cooling load. The general endorsement from the breakout group is that the design of facilities should begin with calculations of the cooling load posed by the intended use of a room, and that the efficiency of ventilation depends not only on the rate but also on the room airflow distribution and the microenvironment of the primary cage, among other factors. One particular gap identified by participants was how to determine cooling load in a room with ventilated caging systems, that is, how much of the load is removed from the room by the exhaust and how much heat is transferred to the room from the cage. Data are needed in this area.

This topic led to a discussion of ventilated caging systems. Approximately 12 systems are commercially available worldwide, and all are different in one or more respects. How can these systems be differentiated or evaluated for use under different use situations? Participants discussed a need for guidance on selection of ventilated caging systems based on criteria such as airflow balance to individual cages; airflow distribution within cages; ammonia levels; filtration expectations (e.g., control of particulates); temperature and humidity; containment (negative vs. positive pressure); noise; vibration; exhaust choices; and ergonomics. The group felt in general that standardized test methods for these and other parameters are needed.

Dinner Speaker

Kay E. Holekamp

A View from the Field:
What the Lives of Wild Animals
Can Teach Us About Care of
Laboratory Animals

Kay E. Holekamp

My goal in this presentation is to review briefly a few seminal contributions from classical ethology and contemporary behavioral ecology that might help us develop better guidelines for use and care of laboratory animals. All of these contributions emphasize the importance of understanding the lives of animals in nature as we try to improve laboratory guidelines. I shall illustrate some of my points here with examples drawn from the lives of my own study animal, the spotted hyena (*Crocuta crocuta*), and other free-living mammals.

In his charming treatise on animal behavior titled "A Stroll Through the Worlds of Animals and Men," Jacob Von Uexkull (1934) observed that animals perceive only limited portions of their total environment. He asked the reader to consider a tick perched on a blade of grass, being bombarded at any given moment by thousands of wavelengths of both light and sound, hundreds of thousands of odorant molecules, myriad tactile stimuli, and information regarding gravity, humidity, and ambient temperature. Of all these countless stimuli hitting the tick, only a tiny few are important for its survival and reproduction, and it is only those few stimuli that the tick must sense and to which it must respond appropriately. All other stimuli are tuned out. Von Uexkull called the array of stimuli existing in the sensory-perceptual world of any animal its *Umwelt*. We now understand that the *Umwelt* of each species is unique, and it is important that we understand the *Umwelt* of each species in our care in the laboratory. This allows us to determine what is and is not salient to the animal, which in turn allows us to make regulatory decisions that are truly in the best interests of our animal charges.

A second important contribution from classical ethology is the ethogram. An ethogram is a complete descriptive inventory of an animal's normal behavior, and as Niko Tinbergen (1951) pointed out, development of an ethogram is a critical first step in attempting to understand the behavior patterns of any species. All behaviors in the ethogram should ideally be described strictly in terms of the animal's motor patterns and the contexts in which they occur, with minimal interpretation by the observer. Familiarity with an animal's complete behavioral repertoire can be very useful to those of us working in the laboratory, because this information allows us to identify pathological behavior, distress, and contentment in our animal charges. New behaviors that arise, or normal behaviors that vanish from the ethogram, usually signal that something is wrong with the living conditions we have made available to the animal.

The third contribution I shall briefly describe from classical ethology is a systematic comparison of the behavior of wild and captive savannah baboons performed by Thelma Rowell (1967). After intensive observations of wild baboons in Africa, she examined the behavior of a troop of conspecific baboons maintained in captivity. Although her captive troop was housed in a large seminatural enclosure, and although the inventory of behaviors emitted by Rowell's baboons was the same in captivity and the wild, the rates at which certain behaviors occurred differed dramatically. Specifically, she found that rates of all social interactions were four times higher in captivity than in the wild, and that rates of aggression were eight times higher in captivity. These escalated rates of behavior presumably occurred because the captive animals had fewer opportunities to resolve their conflicts by moving away from each other. It is highly useful for us to understand conditions of life in nature for any species, and how these differ from conditions in the laboratory, because this understanding allows us to modify the captive environment in the best interests of the captive animals. Assuming we seek to maximize the external validity of the work we perform with captive animals, the more natural the animal's behavior and physiology, the better our science.

In recent decades, myriad studies in behavioral ecology have taught us that animals in nature confront multiple selection pressures every single day of their lives. To survive in the wild, animals must cope effectively with bad weather, hunger, thirst, intra- and interspecific competitors, predators, parasites, and pathogens. Furthermore, these selection pressures often act on animals in opposing directions, such that animals are forced to make trade-offs (e.g., Stearns 1992). For example, avoiding predation may be easier if one's body size is larger, whereas ingesting enough calories to remain well fed may be easier if body size is smaller.

Thus, over the course of evolutionary time, the animal may become a bit larger than optimal for feeding itself, but a bit smaller than optimal for purposes of evading predators. To accommodate opposing selection pressures, wild animals also routinely make shorter-term trade-offs. For instance, even though a rodent may be extremely hungry as it sets off to forage from its home burrow, the animal will nevertheless forego feeding altogether for a while longer if it detects a predator lurking outside the burrow entrance.

When we bring animals into the laboratory, we expose them to a suite of artificial selection pressures that are quite different from the selection pressures they would encounter in nature. As their caretakers, we must make the same sorts of trade-off decisions for them that the animals would make on their own to accommodate opposing selection pressures in nature. Although we cannot ask animals directly about their preferences, these can often be inferred from the animals' behavior or physiology. For example, in our free-living spotted hyena subjects in Kenya, we have gathered preliminary behavioral data indicating that although the hyenas are not terribly bothered by the intramuscular injection involved in being hit by a dart during routine immobilizations, they find the experience of anesthesia itself to be utterly terrifying. In addition, plasma glucocorticoid levels of hyenas that took only a few minutes longer than average to become unconscious were several times as high as in hyenas for which time to unconsciousness was in the normal range (8-13 minutes). Thus both the physiology and the behavior of the hyenas suggest that they find the experience of being out of control of their own bodies extremely stressful.

Robert Sapolsky has observed the same phenomena in his free-living baboons in Kenya (Sapolosky 1982). Noninvasive methods are of course always preferable to invasive procedures; but where a choice must be made between use of anesthesia and causing our animals momentary pain or discomfort in our research, we must be careful to ensure that the trade-off decisions we make on behalf of our laboratory animals are truly based on the best interests of those animals. If common laboratory animals respond to being anesthetized in the same way as do our hyenas or Sapolsky's baboons, then momentary pain or stress might often be vastly preferable, from the animal's point of view, to being anesthetized for minor procedures.

Finally, contemporary behavioral ecology has taught us that variation in nature is enormous and that free-living animals therefore inevitably confront a range of conditions rather than just one "average" condition. Everyone who works with laboratory animals recognizes that it would be inappropriate to apply identical husbandry practices to groups of zebra, zebra finch, and zebra fish. Yet, in addition to obvious interspecific differences, a great deal of variation often exists among free-living members of a single species. In the wild, conspecifics vary among populations occurring in different habitat types. For example, mice in the genus *Peromyscus* breed throughout the North American continent, across a huge latitudinal gradient ranging from 15° N to 60° N (Bronson 1989). It would therefore be very difficult to select one set of environmental conditions to which *Peromyscus* are exposed in their breeding range and declare that set to be the "typical" set of conditions encountered in nature by this species.

Similarly, spotted hyenas occur in virtually all habitat types in sub-Saharan Africa, including the arid sands of the Kalahari Desert, the watery world of the Okavango Delta, the dense forests of central and western Africa, and the prey-rich short-grass plains of the Serengeti ecosystem. How then would one describe the environmental conditions confronted in nature by the "average" spotted hyena? Long-term field work on East African (Kruuk 1972) and Kalahari (Mills 1990) hyenas has shown that body size, diet, home range size, social group size, and circadian activity all differ significantly between these two habitats. Interestingly, however, two things remain constant between habitats: the basic structure of hyena society and the behaviors occurring in the species' ethogram developed at each site.

Conspecific animals vary not only among habitat types but also among populations occupying a single habitat. For example, reproduction and behavior vary quite dramatically between spotted hyena populations separated by only 60 kilometers within the Serengeti ecosystem in eastern Africa. Long-term study of hyenas in the southern part of this ecosystem by Hofer and East (1993a,b,c; 1995) has shown that prey numbers available to the resident hyenas in this area vary enormously with season of the year. During times of the year when prey are scarce, the resident hyenas must commute long distances to feed. They therefore have huge home ranges that encompass their commuting routes, and their attendance at dens to care for their cubs is sporadic. By contrast, in the northern portion of this same ecosystem, Dr. Smale, our colleagues, and I have found that prey are available year-round to resident hyenas, hyenas feed within relatively small home ranges rather than commute, and they attend their cubs at dens daily (Boydston and others 2003; Cooper and others 1999; Holekamp and others 1997b).

Finally, even within a single population, variation in the conditions animals confront may be surprisingly large. Variation among individuals within a single wild population often has a number of different sources including age, sex, social rank, dispersal status, and reproductive condition (e.g., Boydston and others 2001; Holekamp and Smale 1998; Szykman and others 2001). For example, variables that vary dramatically with social rank among free-living spotted hyenas include age at first reproduction, reproductive success, family size, home range size, patterns of association with conspecifics, and even parasite load (Boydston and others 2003; Engh and others 2003; Holekamp and others 1996, 1997a).

Most scientists work on animals in the laboratory rather than in the wild precisely to minimize the kind of variation I have described herein. Then why worry about it? My response is that naturally occurring variation is important to those regulating laboratory animal care because this variation suggests that even for a single species there is often likely to be an entire range of conditions under which the species will thrive in the laboratory.

In summary, classical ethology and modern behavioral ecology have taught us that every animal comes into the laboratory with an evolutionary past and a set of traits shaped by natural selection. These include an *Umwelt*, a normal repertoire of behaviors, and an ability to survive and reproduce under a range of conditions. These traits should factor into our decision making about laboratory animal care guidelines. Given the diversity of conditions under which most species exist in nature, it seems reasonable to expect that a heterogeneous array of husbandry conditions can be utilized in the laboratory without compromising our ability to maintain a homogeneous set of ethical standards for the treatment of these animals.

REFERENCES

Boydston, E.E., Morelli, T.L., Holekamp, K.E. 2001. Sex differences in territorial behavior exhibited by the spotted hyena (*Crocuta crocuta*). Ethology 107:369-385.

Boydston, E.E., Kapheim, K.M., Szykman, M., Holekamp, K.E. 2003. Individual variation in space utilization by female spotted hyenas (*Crocuta crocuta*). J Mamm 84:1006-1018.

Bronson, F.H. 1989. Mammalian Reproductive Biology. Chicago: University of Chicago Press.

Cooper, S.M., Holekamp, K.E., Smale, L. 1999. A seasonal feast: Long-term analysis of feeding behavior in the spotted hyaena Crocuta crocuta (Erxleben). Afr J Ecol 37:149-160.

Engh, A.L., Nelson, K.G., Peebles, C.R., Hernandez, A.D., Hubbard, K.K., Holekamp. K.E. 2003. A coprological survey of the parasites of spotted hyaenas (*Crocuta crocuta*) in the Masai Mara National Reserve, Kenya. J Wildl Dis 39:224-227.

Hofer, H., and East, M.L. 1993a. The commuting system of Serengeti spotted hyaenas: How a predator copes with migratory prey. I. Social organization. Anim Behav 46:547-557.

Hofer, H., and East, M.L. 1993b. The commuting system of Serengeti spotted hyaenas: How a predator copes with migratory prey. II. intrusion pressure and intruders' space use. Anim Behav 46:559-574.

Hofer, H., and East, M.L. 1993c. The commuting system of Serengeti spotted hyaenas: How a predator copes with migratory prey. III. Attendance and maternal care. Anim Behav 46:575-589.

Hofer, H., and East, M.L. 1995. Population dynamics, population size, and the commuting system of Serengeti spotted hyaenas. In: A.R.E. Sinclair and P. Arcese, eds. Serengeti II: Dynamics, Management, and Conservation of an Ecosystem. Chicago: University of Chicago Press. p. 332-363.

Holekamp, K.E., and L. Smale. 1998. Dispersal status influences hormones and behavior in the male spotted hyena. Hormones Behav 33:205-216.

Holekamp, K.E., Smale, L., Szykman, M. 1996. Rank and reproduction in the female spotted hyaena. J Reprod Fertil 108:229-237.

Holekamp, K.E., Cooper, S.M., Katona, C.I., Berry, N.A., Frank, L.G., Smale, L. 1997a. Patterns of association among female spotted hyenas (*Crocuta crocuta*). J Mamm 78:55-64.

Holekamp, K.E., Smale, L., Berg, R., Cooper, S.M. 1997b. Hunting rates and hunting success in the spotted hyaena. J Zool Lond 242:1-15.

Holekamp, K.E., Szykman, M., Boydston, E.E., Smale, L. 1999. Association of seasonal reproductive patterns with changing food availability in an equatorial carnivore. J Reprod Fertil 116:87-93.

Kruuk, H. 1972. The Spotted Hyena. Chicago: University of Chicago Press.

Mills, M.G.L. 1990. Kalahari Hyaenas. London: Unwin-Hyman Ltd.

Rowell, T.E. 1967. A quantitative comparison of the behaviour of a wild and caged baboon group. Anim Behav 15:499-509.

Sapolosky, R.M. 1982. The endocrine stress response and social status in the wild baboon. Hormones Behav 16:279-292.

Stearns, S. 1992. The Evolution of Life Histories. Oxford, UK: Oxford University Press.

Szykman, M., Engh, A.L., Van Horn, R.C., Funk, S., Scribner, K.T., Holekamp, K.E. 2001. Association patterns between male and female spotted hyenas reflect male mate choice. Behav Ecol Sociobiol 50:231-238.

Tinbergen, N. 1951. The Study of Instinct. Oxford, UK: Oxford University Press.

Uexkull, J. Von. 1934. A stroll through the worlds of animals and men. In: C.H. Schiller, ed. Instinctive Behavior. New York: International Universities Press. p. 5-80.

Session 5

Environmental
Enrichment Issues

Enriching the Housing of the Laboratory Rodent: How Might It Affect Research Outcomes?

William T. Greenough and Ann Benefiel

This presentation focuses on the research cost-benefit aspect of enrichment of housing conditions for laboratory rats and mice. The choice of this subject emerged because the session organizers requested a presentation on "laboratory animal housing enrichment" and included the following in their letter of charge:

> The workshop will . . . focus . . . on identifying gaps in the current knowledge in order to encourage future research endeavors, assessing potential financial and outcome costs of unscientifically-based regulations on facilities and research, and determining possible negative impacts of arbitrary regulations on animal welfare.

The basic view put forth herein is that caution is warranted in the adoption of environmental enrichment procedures, because they may complicate interpretation of research results.

We begin by briefly discussing the history of what has come to be called "enriched housing." The first description of an effect of enhanced living conditions on behavior as an indication of altered brain function was the work of Hebb (1949), who compared rats that he reared as "pets" in his home with counterparts reared under normal laboratory conditions (an experiment unlikely to be repeated, given current regulations regarding research animal housing!). Hebb reported that the home-reared rats were superior to the laboratory rats on complex problem-solving tasks and that they continued to move ahead as they were tested on successive

177

tasks. Subsequently, students of Hebb or others inspired by him repeated the basic finding (in the laboratory), that a more stimulating rearing environment enhanced performance on complex learning tasks (e.g., Bingham and Griffiths 1952; Forgays and Read 1962).

Subsequently, Krech and colleagues (1960) reported effects of a similar rat housing environment, for which they adopted the term "enriched," on measures of the activities of enzymes involved in metabolism relating to cholinergic synaptic transmission. This program led to the discovery that some regions of the cerebral cortex were actually heavier and thicker in the "enriched condition" (EC) rats compared with "impoverished condition" (IC) rats kept in barren individual cages (Diamond and others 1966). Research stimulated by theirs triggered the first report of altered dendritic branching (Holloway 1966), although that paper used methods that were inadequate to quantify dendritic branching.

Research on the details of changes induced by such experiences, and that of a number of others, was inspired by the work of Rosenzweig and colleagues. For purposes of this illustration, the work of the Greenough laboratory is selectively emphasized here. An early replication of the Holloway (1966) study using quantitative methods (Volkmar and Greenough 1972) indicated that the dendritic branching of neurons in the rat visual cortex was altered in EC versus IC rats and that "social condition" rats housed in pairs in standard laboratory cages (SC) were intermediate, often differing statistically from both EC and IC rats. This latter result suggests that the more minimalist rodent enrichment procedures such as social housing, which are common in European laboratories and becoming more so in laboratories in the United States now, may actually bring about subtle but detectable changes in the brain. The enriched environment used by the Greenough laboratory, although likely falling short of Hebb's home is, by contrast, a very complex arrangement of objects for play and exploration as Figure 1 indicates.

The effects of these different environments are not restricted to the brain and to the behavior it enables. Significant peripheral somatic differences exist between rats housed in EC and those in IC, which could interact with various sorts of treatments or affect responses to edible reinforcements (Black and others 1989). These differences include, in rats in our laboratory, (1) greater body weight in IC than in EC rats, accompanied by (2) greater food consumption in the ICs, (3) more rapid maturation of the long bones in IC versus EC rats, (4) sometimes greater adrenal to body weight ratios in EC versus IC rats, (5) a higher kidney to body weight ratio in the EC group, and (6) a lower thymus to body weight ratio in the ECs (with no indication of diminished EC immune competence; Black and others 1989). The fact that the organ weight ratios differ in both directions (EC > IC and IC > EC) suggests that they do not reflect merely

FIGURE 1 Enriched rat cage.

the relatively lower body weights of the ECs. The goal here is not to try to explain why these differences occur, but rather to illustrate that a variety of experimental measurements could be affected by differences of this sort. Hence, on the basis of peripheral measures alone, utilizing rats exposed to enriched housing in other experiments could generate confounded results, and certainly switching from nonenriched to enriched animals could generate changes in experimental outcomes. It should also be noted that male and female rats can differ in their responses to enriched environments (e.g., Juraska 1991, 1998). Thus, basic somatic physiological processes are affected by rearing environment complexity, which *can* affect research outcomes if those processes *are* or *affect* variables of interest, and caution is warranted in introducing novel degrees of environment complexity or "enrichment" into ongoing research paradigms.

The brain effects of EC are even more profound. Neurons and their synapses, vasculature, and the two most prominent types of glial cells all are dramatically affected by exposure to an enriched environment. In visual cortex the number of synapses per neuron is 20 to 25% greater in EC rats compared with those in IC, with rats socially housed in cages typically little different from individually housed rats (Turner and

Greenough 1985). This effect is supported by equally substantial increases in the size of the dendritic fields of neurons (Volkmar and Greenough 1972). Synapse morphology and architecture are also different in EC versus IC rats (Jones and others 1997; West and Greenough 1972).

The volume of capillary per neuron is similarly selectively increased in EC rats (Black and others 1987), presumably in part to "power" the increased numbers of synapses, a substantial fraction of which are closely associated with mitochondria on the presynaptic side. Astrocytes, which serve to optimize many metabolic functions of neurons, and can be identified by the presence of their characteristic glial fibrillary acidic protein, are increased in both size and number in EC rats (Sirevaag and Greenough 1991). Moreover, synapses in EC rats are more completely covered by fine astrocytic processes than in IC rats (Jones and Greenough 1996). The other macroglial cell type, the oligodendrocyte that gives rise to the axonal myelination that enhances the speed of conduction of nerve impulses, is also affected by environment enrichment: EC rats have more myelinated axons in the corpus callosum than IC rats (Juraska and Kopcik 1988). All of the foregoing findings have been demonstrated in visual cortex (or connecting callosum), and many effects have also been demonstrated in other brain regions. Taken as a whole, these results indicate that the properties of most cell types and the ways in which they relate to each other in the brain may be altered by the housing environment.

Most of these effects also occur in rats put into enriched environments for the first time as adults. Most rats used in research are purchased from suppliers as young adults, typically shortly after they reach the point of sexual maturity, and are used as quickly as possible after they have become accommodated to their new surroundings, in an effort to minimize cost. If the rats were made to accommodate to an enriched laboratory environment, their bodies and brains might be in a state of relative physiological and structural turbulence at just the time they were expected to be ready to participate in experiments. Clearly research on or involving these variables will be affected, and research on other interacting variables might also be affected in unpredictable ways.

The mechanisms mediating these effects are largely unknown. Neurotrophic factors such as "brain-derived neurotrophic factor" are known to be altered by environmental variations such as enrichment and exercise (Klintsova and others, submitted; Oliff and others 1998; A.Y. Klintsova, E. Dickson, R. Yoshida, and W.T. Greenough, manuscript in preparation), and these factors may well be a part of the process that generates the responses in brain physiology and structure in response to altered environmental conditions. This further complicates the stability of the background against which experimental effects are to be measured.

A novel finding that can be discussed only after it is accepted for publication is Richard Smeyne's finding that the drug MPTP, which kills catecholamine neurons in the substantia nigra in humans and in conventionally housed rats, does not result in similar neural damage when rats are housed in an enriched environment. Although this finding, of course, suggests important therapeutic directions, it also illustrates the complications that might be induced in a well-developed paradigm by the sudden insertion of enriched housing procedures.

Numerous institutions, including my own, have recommended or even mandated enrichment procedures for laboratory rodents recently. In the case of my university, where group housing or the insertion of novel objects into the cage is mandated, a nonscientific poll of principal investigators using rodents found that only I was aware of this policy, despite the fact that those procedures were being applied to their animals at the time I asked them the question. Certainly they did not seem to have been asked whether they thought these procedures might interfere with their research, despite a clear policy guideline with regard to the following statement: "Investigators who must singly cage animals and feel that enrichment materials may confound their research objectives must provide justification." Enrichment appears to have been accepted as a "good thing," with little consideration of its possible effects on experimental outcomes. Taken literally, this University of Illinois policy might make it difficult to determine effects of enrichment that one did not know to exist. There is also tacit acceptance that group housing is superior to individual housing despite data that call into serious question whether this is true (Bartolomucci and others 2003).

Several presentations at the current meeting seemed similarly to espouse such a view. Perhaps most disconcerting is the arbitrary assumption that enrichment is better for the animals, with little data to support this assumption beyond the fact that the animals attend to enrichment objects and appear to play more vigorously when such objects are present. It appears in this case, as in the case of several other presentations at this workshop, that the animals' preferences are being allowed to drive, if not dictate, the issue of what constitutes enrichment. In this regard, it is of value to note that animals' preferences may not be the ideal guideline to what is of most value to them. In earlier research on addiction, which would probably not be permitted today, it was found that rats and monkeys given unrestricted or nearly unrestricted access to drugs of abuse (cocaine, amphetamine, methamphetamine, and alcohol) would self-administer these drugs within 1 month to the point of cessation of eating, refusal of hand-fed treats, and in many cases until dead, or near enough to death that researchers removed them from the experiment and provided life-saving measures to keep them alive (e.g., Johanson and others 1976; Pickens and Thompson 1971). This and similar findings in other self-selection domains suggests that the animals' judgments are not always in synchrony with what appears optimal to their health (e.g., Galef and Beck 1990).

Thus we propose the following recommendation regarding the sudden and arbitrary insertion of environmental enrichment procedures into ongoing research: **Caution is warranted.** We should not assume that enrichment will not affect our research measurements or outcomes unless demonstrated otherwise. And we should not mandate enrichment of animals engaged (or to be engaged) in a research protocol unless the protocol has been explicitly shown not to be affected by the enrichment procedure to be used (or the effects are known and taken into account). Finally, just as we should use caution in generalizing from humans to mice about what we believe is best for a mouse or a human (see "To a mouse . . ." by R. Burns [Douglas 1993]), we should also use caution when we generalize across more closely related species, until an experimental basis for doing so has been established.

> Still thou art blest, compar'd wi' me
> The present only toucheth thee:
> But, Och! I backward cast my e'e.
> On prospects drear!
> An' forward, tho' I canna see,
> I guess an' fear!

> And finally,
> The best-laid schemes o' mice an' men
> Gang aft agley
> —Robert Burns

REFERENCES

Bartolomucci, A., Palanza, P., Sacerdote, P., Ceresini, G., Chirieleison, A., Panerai, A.E., Parmigiani, S. 2003. Individual housing induces altered immuno-endocrine responses to psychological stress in male mice. Psychoneuroendocrinology 28:540-558.

Bingham, W.E., and W.J. Griffiths, Jr. 1952. The effect of different environments during infancy on adult behavior in the rat. J Comp Physiol Psychol 45:307-312.

Black, J.E., Sirevaag, A.M., Greenough, W.T. 1987. Complex experience promotes capillary formation in young rat visual cortex. Neurosci Lett 83:351-355.

Black, J.E., Sirevaag, A.M., Wallace, C.S., Savin, M.H., Greenough, W.T. 1989. Effects of complex experience on somatic growth and organ development in rats. Dev Psychobiol 22:727-752.

Diamond, M.C., Law, F., Rhodes, H., Lindner, B., Rosenzweig, M.R., Krech, D., Bennett, E.L. 1966. Increases in cortical depth and glia numbers in rats subjected to enriched environment. J Comp Neurol 128:117-125.

Douglas, W.S., ed. 1933. Collected Works of Robert Burns. London: Taylor & Francis Books Ltd.

Forgays, D.G., and J.M. Read. 1962. Crucial periods for free-environmental experience in the rat. J Comp Physiol Psychol 55:816-818.

Galef, B.G., and M. Beck. 1990. Diet selection and poison avoidance by mammals individually and in social groups. In: Handbook of Behavioral Neurobiology, Vol. 10, Neurobiology of Food and Fluid Intake. p. 329-349.

Hebb, D.O. 1949. The Organization of Behavior. New York: John Wiley & Sons.

Holloway, R.L. 1966. Dendritic branching: Some preliminary results of training and complexity in rat visual cortex. Brain Res 2:393-396.

Johanson, C.E., Balster, R.L., Bonese, K. 1976. Self-administration of psychomotor stimulant drugs: The effects of unlimited access. Pharmacol Biochem Behav 4:45-51.

Jones, T.A., and W.T. Greenough. 1996. Ultrastructural evidence for increased contact between astrocytes and synapses in rats reared in a complex environment. Neurobiol Learning Mem 65:48-56.

Jones, T.A., Klintsova, A.Y., Kilman, V.L., Sirevaag, A.M., Greenough, W.T. 1997. Induction of multiple synapses by experience in the visual cortex of adult rats. Neurobiol Learning Mem 68:13-20.

Juraska, J.M. 1991. Sex differences in "cognitive" regions of the rat brain. Psychoneuroendocrinology 16:105-119.

Juraska, J.M. 1998. Neural plasticity and the development of sex differences. Ann Rev Sex Res IX:20-38.

Juraska, J.M., and J.R. Kopcik. 1988. Sex and environmental influences on the size and ultrastructure of the rat corpus callosum. Brain Res 450:1-8.

Krech, D., Rosenzweig, M.R., Bennett, E.L. 1960. Effects of environmental complexity and training on brain chemistry. J Comp Physiol Psychol 53:509-514.

Oliff, H., Berchtold, N., Isackson, P., Cotman, C. 1998. Exercise-induced regulation of brain-derived neurotrophic factor (BDNF) transcripts in the rat hippocampus. Mol Cell Res 61:147-153.

Pickens, R., and T. Thompson. 1971. Characteristics of stimulant drug reinforcement. In: T. Thompson, ed. Stimulus Properties of Drugs. New York: Appleton-Century-Crofts. p. 172-192.

Sirevaag, A.M., and W.T. Greenough. 1991. Plasticity of GFAP-immunoreactive astrocyte size and number in visual cortex of rats reared in complex environments. Brain Res 540:273-278.

Turner, A.M., and W.T. Greenough. 1985. Differential rearing effects on rat visual cortex synapses. I. Synaptic and neuronal density and synapses per neuron. Brain Res 329:195-203.

Volkmar, F.R., and W.T. Greenough. 1972. Rearing complexity affects branching of dendrites in the visual cortex of the rat. Science 176:1445-1447.

West, R.W., and W.T. Greenough. 1972. Effect of environmental complexity on cortical synapses of rats: Preliminary results. Behav Biol 7:279-284.

Search for Optimal Enrichment

Timo Nevalainen

Recently a Council of Europe (CoE) expert group emphasized the need for environmental enrichment and group housing as *refinements* for all laboratory species unless there are scientific or veterinary reasons not to do so (Hansen and others 1999). This emphasis is part of the revision of CoE Appendix A, in which species-specific recommendations serve as a starting point in the choice of enrichment.

Indeed, the questions are not *whether to use* but *how to use* enrichment, and *how far* it should be regulated. Enrichment as such is unfortunately a poorly defined entity, which can be considered to include, for example, group housing, a variety of added items into cages, and even bedding. If this wide definition is accepted, we all will be using enrichment. This variety and the fact that most of us practice our own enrichment make it very difficult or even impossible to draw general conclusions on the effects of enrichment, which adds to the confusion. Overall, the situation is partly out of hand, and corrective action is desperately needed.

We often refer to harmonization as the ultimate goal of international cooperation. How does harmonization relate to enrichment requirements, ethics, and science? In this context, environmental enrichment should be seen as the minimum standard, below which no one is allowed to operate. Well above the minimum standard, there should be an area of excellence, where ideals of ethics and science are the driving forces. Any refinement in housing to improve animal welfare requires:

- scientific validation
- that the refinement is truly beneficial for the animals (efficacy)
- that the refinement does not detract from the scientific integrity (safety).

In other words, one should always ask whether a refinement or enrichment has value and whether it hurts science. These factors must be seen as the key criteria of optimal enrichment. The main emphasis of the European guidelines, as well as other guidelines, is on animal welfare, and much less on safeguarding scientific integrity. But how can one compare results from experiments performed in different laboratories when variable environmental enrichment strategies are used throughout the world?

Three different approaches to enrichment strategies can be seen. Some people think that anything can be used, and they have long lists of items of various origins. Others practice a precautionary principle with items that have no or poor scientific basis. They try to determine whether there is a potential danger of scientific interference and then act accordingly. A third approach is to use only the enrichment items based on scientific evidence to show both efficacy and safety. This last approach is considered too cynical by others.

Enrichment has been shown to change the animal's behavior and physiology, which are indeed the main goals of the practice. But this alteration also means that the animals are not the same as those used in earlier experiments. This result gives rise to the concern that the scientific data from earlier studies may—at least partly—have become useless.

The revision of Appendix A states that enrichment may be omitted if there is a welfare, veterinary, or scientific reason to do so. Interference with an experimental outcome could be an example of a scientific reason and fighting between incompatible animals a veterinary reason. Inappropriate enrichment may result in mortality, morbidity, aggression, and overt stress. All of these results are expressions of a compromised welfare.

Selection of materials to be used as enrichment in cages is critical, because some substances may cause interference with well-being and experimental results. Bedding made from soft wood can contain high concentrations of volatile compounds, especially α- and β-pinenes, causing major changes in liver microsomal enzyme activity. A study published 7 years ago showed that even then many commonly used types of bedding contained many organic volatile compounds, and that autoclaving decreased concentrations to a fraction of the original values (Nevalainen and Vartiainen 1996).

If enrichment items are made of organic materials, they should meet the same chemical criteria as bedding. When we compared 15 beddings with 16 enrichment items, we found that the range and concentrations of organic volatile compounds were quite similar, but there were a few bedding and enrichment items with unwanted pinene components (Meller and others 2004). The obvious solution here would be to use enrichment items and bedding that are of the same material in an effort to decrease the chemical burden on the animals. Unfortunately, only a few such products are available.

If this "no new material into the cage approach" cannot be used, then the next best approach is to use inert materials. Earlier presentations in this meeting have shown that some plastic materials that are commonly considered inert, such as polycarbonate and PVC, emit compounds that have a considerable impact on physiology of the animals. Chemical aspects are usually brought up in good laboratory practice (GLP) studies, but they are equally real in non-GLP studies. The question is not whether chemical exposure may lead to toxicity, but whether it may have the potential to change study results.

Some people advocate rotation of enrichment. At least in rodents and rabbits, it does not seem like a good idea to change enrichment, particularly within a study. If the necessity of rotation is reasoned with novelty, one should consider looking for enrichment with longer-lasting value. For instance, combining diet and enrichment is something commonly practiced with nonhuman primates but almost untouched with rodents and rabbits.

Modification to the environment could interfere with a study in two ways: (1) it may change the mean, but it should do so in all groups, or (2) it may change the variance, and this change is bound to increase the number of animals used. However, the opposite may happen and results could be improved, leading to fewer animals being used. Indeed, instead of trying to assess the impact of enrichment items on every possible determination, it might be more productive to look at the effects on variation of welfare indicators, with the understanding that low variation here is likely to show as low variation in most other determinations. At the same time, it is important to achieve the most uniform welfare of the animals in the study. Indeed, better science may evolve from the application of the two Rs, refinement and reduction, depending on the precise nature and purpose of the experiment.

"Minimum floor area for one or two socially harmonious rabbits" in a space allocation table is a new approach suggested by the CoE expert group on rodents and rabbits. The minimum space required for rabbits increases considerably at the same time, but this enrichment "catch" means that facilities can house the same number of rabbits, if pair housed, in new cages than single housed in old cages (see Table 1).

We carefully followed the minimum specifications of the expert group for rabbits and even exceeded the specifications to an extent in space

TABLE 1 Comparison of Minimum Space Requirements for Stock Rabbits: Council of Europe 1986 vs. 2000

	CoE ETS 123, 1986, Appendix A			CoE ETS 123. 2000, Appendix A		
	When housed singly				For 1 or 2 socially housed animals	
Weight	Floor Area (cm²)	Height (cm)		Weight	Floor Area (cm²)	Height (cm)
Up to 2 kg	1400	30				
Up to 3 kg	2000	30		< 3 kg	3500	45
Up to 4 kg	2500	35				
Up to 5 kg	3000	40		3-5 kg	4200	45
Above 5 kg	3600	40		> 5 kg	5400	60

provided, where a shelf would not have been necessary. We compared sister pairs housed either in pairs or singly with four sampling periods over a 144-day study using six different serum chemistry parameters. Statistical analysis showed that growth and alkaline phosphatase showed less variance in pair-housed animals. When this result is calculated into the number of animals needed, the shift from pair-housed to individually housed animals would, according to statistical power analysis, mean using multipliers 1.4 and 2.4, correspondingly (Nevalainen and others 2003).

Because each species has different behavioral and physiological needs, one standard enrichment program for all species is unrealistic. To perform enrichment responsibly, the process must be species specific as defined in the Appendix A revision. To refine housing, we can conduct preference tests and economic demand tests on other types of proposed enrichments wherein animals have to work in some way to receive the "refinement." This approach measures their motivation against other "needs" (e.g., food or a social partner) and helps us understand what animals consider important.

The assessment of refinement requires scientific methods that do not disturb the animals, and evidence of normal values is one indicator of the impact of a refinement. With one or more efficient welfare indicators, it is possible to recognize and assess refinement aspects of both animal housing and experimental procedures. Promotion of animal well-being, as seen by a reduction in disease or damaging behaviors, is important when assessing housing environment.

The Federation of European Laboratory Animal Science Associations (FELASA) has established a Working Group on Standardization of

Enrichment for rodents and rabbits (FELASA 2003). The terms of references for the group read as follows: "How to standardize enrichment in rodents and rabbits with essential species-specific needs, needs of gender and life stage and animal welfare (defined as functioning and feeling well) are guaranteed and interference with studies minimized."

We may try to start with a commonsense approach: What are the things that should *not* be done with enrichment? One should change enrichment as seldom as possible, like one would change the type of diet and bedding, and *never* within a study. One should prefer a truly inert or a no-new-materials practice. Perhaps we should aim at an "enrichment profile," starting from the breeder and continuing to the completion of the experiment. Furthermore, one should look for changes in the variance of results and for deleterious effects (e.g., fighting) on animal welfare. Standardization in this context means that there should be a limited number of standardized, efficient, and safe species-specific enrichments. Obviously, this task is challenging and much easier said than done.

Best practice should be based on scientific data and aim well beyond harmonization. Regulations, which may be difficult to update regularly, should leave space for adjustments in best practice. There are urgent legal, ethical, and scientific expectations for guidelines on optimal and standardized enrichment.

REFERENCES

FELASA [Federation of European Laboratory Animal Science Associations]. 2003. FELASA Working Group on Standardization of Enrichment (http://www.felasa.org/working/stenr.html).

Hansen, A.K., Baumans, V., Elliot, H., Francis, R., Holgate, B., Hubrecht, R., Jennings, M., Peters, A., Stauffacher, M. 1999. Future principles for housing and care of laboratory rodents and rabbits. Report concerning revision of the Council of Europe Convention ETS 123 Appendix A concerning questions related to rodents and rabbits issued by the Council's working group for rodents and rabbits. Part A, Actions and proposals of the working group.

Meller, A., Laine, O., Voipio, H-M., Vartiainen, T., Nevalainen, T. 2003. Volatile organic compounds in animal bedding and enrichment items. (Abstract.) Presented at the 9th FELASA Symposium held in Nantes, France, June 14-17, 2004.

Nevalainen, T., and T. Vartiainen. 1996. Volatile organic compounds in commonly used beddings before and after autoclaving. Scand J Lab Anim Sci 23:101-104.

Nevalainen, T.O., Nevalainen, J.I., Guhad, F.A., Lang, C.M. 2003. Pair housing of rabbits and variation in serum chemistry. (Abstract.) Presented at the AALAS 2003 National Meeting, Seattle, WA, USA, October 12-16, 2003.

Breakout Session:
Environmental Enrichment Issues:
Mice/Rats/Rabbits

Leaders: John G.Vandenbergh and Vera Baumans

Rapporteurs: Primary, Jennifer Obernier;
Secondary, Stephen W. Barthold

The informal introductory comments of Drs. Baumans and Vandenbergh stimulated immediate discussion. Dr. Baumans discussed the pros and cons of enrichment, emphasizing the necessities for taking into account the normal behavior of each species and for evaluating enrichment methods. Dr. Vandenbergh elaborated on this point noting that guidelines must have a positive strategy; they should identify a scientific basis and measure outcome appropriately; and they should be performance based. This combination of requirements poses larger issues in that it is difficult to define what to measure, what the approach should be, and how to interpret the findings. Cortisol, for example, is not the Holy Grail to indicate the extent of animal welfare. Stress and steroid responses have both good and bad effects, depending on circumstance. Dr. Vandenbergh further indicated that guidelines must not be based on subjective measurements; they must garner respect of the scientific community and must have sensitivity to the needs of science.

Participants felt that guidelines, and the creation of new guidelines, must encourage and stimulate science. Institutional animal care and use committees, for example, could help facilitate science by filling voids in the knowledge base by encouraging needed studies that are specific to their institutions or needs. Rigid regulations or interpretation of guidelines as such tend to place restrictions on process, thus yielding less science-based information. Primate enrichment guidelines are a good example. Enrichment programs are required, but the institution is left to be creative in implementing an enrichment plan. This situation encourages creative approaches in lieu of standardized and intrusive regulations.

Nevertheless, the argument was advanced by some in the group that it may be necessary to establish standards *before* scientific proof, thereby stimulating research. However, the group also emphasized that in the absence of scientific information, standards should be more general, thereby stimulating research leading to more specific standards. There was divided opinion on this subject.

The group discussed sources of funding for developing science-based guidelines. Several options were mentioned, including the American College of Laboratory Animal Medicine Foundation and the Johns Hopkins Center for Alternatives to Animal Testing.

The discussion elucidated the reality of laboratory animal welfare. Clear differences exist between Europe and the United States, as well as Asia. The European approach emphasizes detailed regulations that tend to be inflexible, whereas the US and Canadian approaches are based on general guidelines that encourage new approaches and flexibility. The Japanese approach is more cultural and is based on respect of animals and Buddhist philosophy. There is misunderstanding of the US and European policies, and there is misunderstanding of our own respective systems. Shorthand versions of more complex guidelines and regulations tend to be used. Some Europeans stated that there is public pressure for change, and thus they cannot wait for science. This discussion led to considerable response that such an approach is frightening and does not serve anyone well.

An additional caveat discussed is that species-specific behaviors on which guidelines and regulations are built are also significant variables. Although basic behavior is retained, domestication inbreeding has adapted animals to the research environment, and there is marked strain-related variation among rodents. Therefore, some participants felt that science-based guidelines should take this adaptation into consideration. Transgenic animals create new challenges. It is dangerous to "lump" rodents, particularly different strains of rodents, together.

It was also noted that many things that make animals "happy" are not necessarily good for them. Drug abuse preference or measurements of brain pleasure centers underscore this concept.

In summary, more questions arise than answers. What should we measure? Who should measure? Who should fund the work? How should the work be funded? How can the general scientific community be rallied to assist? There are no easy answers. (For consideration of these questions see the discussion following the Point/Counterpoint session on p. 201.)

Breakout Session:
Environmental Enrichment for Dogs and Cats

Leader: Graham Moore

Rapporteur: Janet Gonder

The participants began by identifying discussion topics. Topics included enrichment beyond exercise; clarification of the use of structures (do they add or subtract from floor space?); recommendations on socialization; acquisition of animals (e.g., experience and socialization at the vendor); vertical space for dogs; exercise (what, when, how, why); and Council of Europe requirements. Most of the discussion was directed at dogs, but specific issues for cats were noted. The term "enrichment" was considered as a complete package to include housing, structures, toys, socialization (with humans and conspecifics), and exercise. Variability was thought to be of benefit.

But what really counts? Some participants posed the thought that human interaction and provision of a cage mate might suffice. Almost everyone agreed that more could be done to socialize/habituate dogs and cats to the laboratory. Early socialization of dogs is critical. Provision of an "enrichment profile" by vendors of purpose-bred dogs and cats was suggested. Of course, such provision would be difficult to achieve with random source animals, as with knowledge of health status, genetics, and so forth.

Participants listed what they thought were the key components of an integrated enrichment program. Components include socialization, exercise, pen/cage structures, and other physical enrichment items.

Socialization was thought to be critical, particularly in the early development period. This component should include socialization with humans and conspecifics, as well as habituation to the laboratory. The participants generally agreed that single housing should be specifically justified based perhaps on the experimental procedure, genetics or breed variation, specific health or husbandry issues, and individual temperament.

Participants did not think the term "exercise" per se to be a particularly helpful guidance. The Council of Europe uses the phrase "physical activity and/or experiencing novel environments."

Participants found the following to be important when considering structures within the enclosure: privacy (may be more important for cats); some control over social interactions; separate areas for different activities; raised platforms; and subdivisions to allow visual stimulation.

Other physical enrichment might include items to allow chewing behavior in dogs; items to be used in social interactions with cagemates; items for play (pseudo-predatory behavior) in cats; and utilization of vertical space (the opportunity to climb for cats).

In summary, participants believed that institutions should have a written program of care. In addition, the use of laboratory dogs and cats should include integration of multiple components, consider factors that meet both social and behavioral needs of the animals, and take into account procedural or protocol requirements.

Breakout Session: Assessment of Nonhuman Primate Enrichment— Science Versus Welfare Concerns

Leader: Carolyn Crockett

Rapporteur: Randall J. Nelson

Participants discussed the questions and topics that appear below.

- **What is the scientific basis or peer-reviewed literature that influences or drives the assessment of enrichment for nonhuman primates (NHPs)?**
  - **What other influences or factors are involved?**
  - **Where are the gaps in our scientific knowledge?**

Participants outlined the following benefits of performance standards for assessing environmental enrichment in NHPs:

- Promote normal behavior: Stimulate a range of normal behaviors, thereby preventing or reducing the development of abnormal behaviors.
- Reduce abnormal behavior: Redirect activities from abnormal to normal; provide outlets for behaviors that might otherwise be self-directed and possibly injurious.
- Reduce stress and associated physiological imbalances: Increase the ability of the animal to cope with potentially stressful laboratory experiences.
- Improve research: By making a healthier research animal (e.g., with normal physiological values), and by reducing subject attrition from development of severe behavior disorders.
- Other possible benefits.

Other reasons for providing environmental enrichment for NHPs include:

- Satisfying public opinion: Providing visible evidence (e.g., enrichment items) that animal welfare concerns with respect to behavioral management are being addressed.
- Complying with laws: In the United States, complying with the animal welfare regulations requiring environmental enhancement plans adequate to promote psychological well-being.
- Motivation of workers (technicians): Fortifying the value of the work and its scientific benefit.
- Other possible reasons.

Participants also discussed the assessment of environmental enhancement for nonhuman primates. Discussion topics included use, other benefits, costs, scientific evidence, research protocol constraints, species considerations, and other considerations, as outlined below.

Considerations for Enrichment (Scientific evidence supporting use or other benefits vs. professional opinions or anecdotal evidence):
- Use; preference
  - Percentage of time budget devoted to use of item
  - Choice; simple preference testing
  - Economic models: Elasticity of demand; change in consumption or usage when made more costly
- Other benefits
  - Facilitating a variety of normal behaviors
  - Reducing abnormal behavior
    - Duration of reduction
    - Generality of reduction (i.e., all or selected undesirable behaviors; e.g., locomotor stereotypy vs. potentially self-injurious behavior)
  - Reducing stress
    - Cortisol
    - Other physiological measures (not yet identified)
    - Behavioral measures of (dis)stress (not yet identified)
  - Other
    - Improving overall health (measures, to be determined)
    - Possible others
- Costs
  - Monetary cost
  - Time cost: implementation, sanitization, etc.
  - Risk: injury, disease transmission;
    - biosafety concerns for personnel

- ○ Rebound effect; increased cost if enrichment conditions are changed; social and physical
- Research protocol constraints
  - ○ Good laboratory practice; toxicology
  - ○ Testing; sampling (leaving home cage, perhaps)
  - ○ Infectious disease
- Species, gender, age considerations, variability
  - ○ Individual differences exist and are noticeable in NHPs
  - ○ Extent to which maladaptive behaviors are caused by environment or inherent in the individual (neurochemical imbalances, organic disease, i.e., need for analgesia in experimental or naturally occurring procedures)
  - ○ Changes in caretakers
- Other considerations (may affect costs)
  - ○ Documentation of benefit (or lack thereof)
    - ♦ Appropriate documentation and person who reviews it
    - ♦ Existence versus benefit
    - ♦ Determination of intra- versus interinstitutional variability
      — Degree to which literature can suffice, especially with individual NHP variations
    - ♦ Communication with others: veterinarians, principal investigators, IACUC members
    - ♦ Novelty (i.e., whether variety within this category is necessary to achieve measurable benefit)
    - ♦ Frequency of providing this category of enrichment to achieve measurable benefit
    - ♦ Other possible factors

Minimum Standards (scientifically or anecdotally based) (Context-dependent variables should be considered):
- Structural enrichment
  - ○ Perches in cages; climbing structures in larger group enclosures
    - ♦ NHPs prefer perches in preference studies, but may take a few days to adapt. "A useful furnishing."
  - ○ Visual barriers: "privacy panels" in cages, barrels, etc., in enclosures
    - ♦ Probably reduces contact aggression and allows withdrawal. However, data are scarce, and most are anecdotal.
  - ○ Other possible factors

- Manipulanda (relatively durable items)
  - ○ "Toys" in cages

- ○ Mirrors on cages; allow control of environment (viewing of conspecifics)
- ○ Determination of whether behavioral (experimental) manipulanda constitute enrichment
- ○ Habituation to items should be avoided and individual variations recognized
- ○ In the "wild," youngsters play quite a bit with toys. More individual differences characterize adults.
- ○ The cost is low; the potential benefit high

- • Simple food treats (not foraging)
  - ○ Produce
  - ○ Other (peanuts, pasta, etc.)

- • Foraging
  - ○ Devices such as puzzle feeders, foraging boards
    - ◆ Complex versus simple
  - ○ Frozen treats or complex items for browsing, which also prolong consumption time
  - ○ Floor substrate (bedding or woodchips) in group rooms
  - ○ Special discussion about foraging
    - ◆ Should tasks performed for food or drink reward count as "foraging" (if "foraging" experiences specifically required by regulations)

- • Special enrichment items (non-food based)
  - ○ Grooming boards (fleece, turf); paint rollers
  - ○ Destructible: paper, cardboard, wood pieces
  - ○ Other possible items

- • Sensory enrichment (visual, auditory, olfactory)
  - ○ Video, television
  - ○ Murals, colorful shower curtains
  - ○ Music, natural sounds
  - ○ Smells, aromatherapy
  - ○ Windows (to outside, to inside corridor, to other animal rooms)
  - ○ Light level; light cycle

- • Spatial
  - ○ Cage size: Participants did not discuss this topic, but instead reached consensus that cage size should be sufficient to accom-

modate all agreed-upon enrichments and to permit normal postural adjustment.
- ° Cage level
    - ♦ Animals prefer to observe from above, but there is not necessarily a physiological difference to accompany this.
- ° Periodic access to larger "activity cage" (frequency, duration)

- Social Contact
  - ° Tactile social contact with conspecifics
    - ♦ Degrees and type of conspecific contact: visual-only, grooming-contact, pair, small group, typical species-specific group, same-sex, opposite sex, ages, full-time, periodic
      - — Age at weaning
      - — Age at first single housing
      - — Proportion of immature developmental stages spent in single housing
  - ° Compatible human caregiver versus same species versus "compatible" species; determination of whether human contact can compensate for individual housing
    - ♦ Structured human contact (training) versus simple contact (e.g., providing treats). Training is beneficial but not a substitute for conspecific contact.
    - ♦ Habituation to caretakers, handlers, experimenters can be beneficial, as can consistency in surroundings. A balance is essential.

Discussion points included the observation that some who conduct enrichment programs at their institutions may not be completely trained in the behavior of one or more species for which they are specifying enrichment programs. Enrichment effects are additive, and it is difficult to examine the pieces in isolation. The whole may be greater than the sum of the parts.

The consensus of the group was that when the science is not available, expert opinion should be used. With regard to who has the expertise, participants stated that it depends on who has the most experience with the individual NHP in question. The team approach is crucial when establishing the best/good practices to be implemented under the institutional and experimental constraints at any given location. In summary, a cage size should be used that is sufficient to accomplish appropriate enrichment and species-specific behaviors.

Point/Counterpoint:

The Cases For and Against Harmonization

Point/Counterpoint:
The Cases For and Against
Harmonization

William S. Stokes

PARTICIPANTS

Michael D. Kastello
Gilles Demers
Jonathan Richmond
John G. Miller
Nelson Garnett
Wim deLeeuw
Derek Forbes
Clément Gauthier
John C. Crabbe
Naoko Kagiyama

DR. STOKES (William S. Stokes, National Institute of Environmental Health Sciences, Research Triangle Park, NC): I want to thank all who have stayed throughout the meeting to this final session, which I think will be well worth your attendance. We will discuss the cases for and against harmonization of laboratory animal guidelines and how we might approach balancing the need for accomplishing high-quality medical research while providing for optimal animal welfare.

First, I would like to comment that I have been very impressed with this conference and the level of the discussions that have been taking place. I think this workshop is moving us in the right direction. I have been involved in international harmonization for safety testing guidelines for the past 10 years. It is not an easy process, but when harmonized guidelines are finally established, it is a good feeling. So if we move in that direction in the area of animal care and use guidelines, I believe it will be, again, potentially challenging, but a rewarding and worthwhile experience.

Second, I am also impressed by the fact that there are nearly 150 people here, from 14 different countries. This attendance truly reflects an international commitment to animal welfare. Clearly, harmonization has the potential to contribute to increased replicability and comparability of research and testing results based on the use of animals. It also enhances the likelihood that there is a benchmark of animal care that is adhered to on an international basis.

In a perfect world, we would all sit down at the table, look at the scientific data, and reach a consensus on good animal care guidelines, based on that science. But as discussed at this meeting, our guidelines have evolved largely based on professional judgment and the ability and need to obtain animals that are healthy and free of injury, and to facilitate high-quality scientific research and testing. However, this workshop has identified data gaps that need to be filled to have a basis not only for future guidelines, but even for our current guidelines. Some issues have been raised about proposing changes, such as increased cost and the potential to introduce confounding factors or variables in our research and testing. For this reason, I believe that we must proceed with caution in harmonizing guidelines.

Nevertheless, we are at a critical crossroads as to how we move forward. You all have probably heard the famous saying of Yogi Berra: "When you come to a fork in the road, take it." Clearly, there is a fork in the road, and each of us is going to take it. However, we need to proceed on an informed basis and be certain that the direction we choose is the correct one.

Prior to this meeting, we asked the panelists to describe their proposed position with regard to harmonization—whether they were for it or against it. We were hoping to have about six for and six against. As you can imagine, being opposed to harmonization does not sound very favorable. It is like not supporting "motherhood and apple pie." In our results, many responders were for harmonization; some were for harmonization with reservations; and at least one responder was against harmonization. Based on these results, I think the discussion today will be lively.

We will use the following format: I will ask each of the panelists to state his/her position in 2 to 3 minutes, and then to comment on how s/he would advance that position (whether it is for harmonization, for harmonization with reservations, or against harmonization). Next, the panelists will be asked to respond to each other's comments, with the opportunity for questions. We then will take questions and comments from the audience. In this session, we will proceed with the participants in the order in which they are listed on the program.

DR. KASTELLO (Michael D. Kastello, Aventis Pharmaceuticals, Bridgewater, NJ): I would like to commend and thank ILAR for organizing this meeting. It has been very interesting and useful to me. I would consider it, in many ways, a first step in a dialogue that I think will persist for a long time.

I am responsible for research animals and animal care and use programs in France, Germany, Japan, and the United States. Harmonization, whatever it might be, is something that seems as if it should be useful to me, with the diverse responsibilities that I have. But this afternoon I am speaking against harmonization.

I do not know the meaning of "harmonization." Is it animal care? Is it animal welfare? Is it laws, regulations, standards? Is it the minimum standard? Is it best practice?

With more than 30 years of experience, I continue to be amazed by what we have learned from animals about biological mechanisms, human and animal diseases, safety and efficacy of medicines and vaccines, and the risks of natural and man-made products. I am also impressed by what we do not know about the animals we use.

It has been stated repeatedly here the last few days that data gaps exist. I think that is a kind phrase. We have *huge* areas for which we have insufficient information. Some individuals in the audience have interpreted that statement as being a reason for not doing something; however, I think it is a reason for doing something.

More importantly, perhaps, than the data gaps are the knowledge gaps. We can ask, is one study enough, or do we need 50? Yet the application of that information is critical. The knowledge gaps are exacerbated, in my mind, by the cultural gaps, traditions, values, laws, and regulations that already exist, as well as religious differences.

This concept of harmonization is seductive, but we have much to do first. In my opinion, standards, guidelines, regulations, and laws must be science based. They must benefit the animals we use. They must not burden research unnecessarily. It is not sufficient to show that a practice we propose is harmful; I think we must show that it is beneficial to the animals.

We can never address all of the issues, conditions, circumstances, species, strains, and so forth. Therefore, whatever to do in moving forward must include the application of performance standards and professional judgment. I am not yet ready for harmonization.

DR. DEMERS (Gilles Demers, International Council for Laboratory Animal Science (ICLAS), Quebec, Canada): My comments are provided on behalf of the ICLAS. I will speak on behalf of all people around the world who are doing research. Some of them come from countries that are not well developed. They are using animals, and they are trying to find ways to do good science. For this reason, they have requested that ICLAS participate in the process of harmonization of guidelines.

The participation of ICLAS is important for them because their countries are different from ours. All of us have come here from countries that have legislation and guidelines in place, yet other countries not represented here have no guidelines and, in some cases, no legislation. They therefore want to follow what is done in Europe, or in America. However, they need documents to empower what they want to do, because they are working for universities where sometimes the concept of triage is not well supported. They seek statements that have been issued by an international scientific community.

In this workshop, we represent that type of community. We are a large group of individuals involved in several scientific organizations. We have the capacity and the power to make statements and to support documents and guidelines. Because they do not constitute legislation, it is possible for us to agree or disagree on a paragraph or a portion of a guideline. Basically, the guidelines exist to help people understand the ethical value of what they are doing. In that sense, I believe that we should express our support by working, through meetings like this one, for harmonization. Nevertheless, ICLAS opposes standardization because ICLAS is in constant liaison with countries and regions that have different cultures, traditions, religion, legislation, and regulations.

A recent example of these differences has been evident during this workshop in discussions with Dr. Kagiyama, who is from Asia and who will speak later. We have spent much time during these 3 days discussing environmental enrichment, although this concept does not really exist in the perspectives of Asian people. Asian scientists are focusing more on good science and the good quality of their animals' health. They want the biological material they use to be perfect. They understand that animals in good health are animals that are free of stress. To avoid stress, they must help the animals deal with their psychological and physiological needs. In this way, they are providing environmental enrichment. They are providing contact bedding, social housing, and visual access to other animals. They may not provide as many devices as are provided in other countries, but they are performing good science.

For these reasons, I believe that we should expand this recent dialogue in the future to include other countries. We must take into consideration that other countries and other regions of the world do not use the same words as we are using. But we are all in favor of harmonization.

DR. RICHMOND (Jonathan Richmond, Home Office, Dundee,

United Kingdom): As a regulator, my position on harmonization is, "yes, but," and the "buts" are as important as the "yes." I think harmonization is defensible only if the objective is to promote the best science and the best animal welfare, not simply to standardize around something. When I say best animal welfare, I mean that which makes the animals feel good, not that which makes a regulator or an IACUC feel good.

I believe we have established that there is not enough evidence for all of these things to be evidence based. But I think in the absence of evidence, they have to at least be rational, defensible, and based on systems that have been tried and are known to work in practice. I think they should always, if possible, be in the form of performance, rather than engineering, standards. However, I think that on many fronts, we are not even ready for the standard. We are probably ready to agree only on the principle around which the standard should be stenciled. If there are performance standards, then I believe there must be clarity about how compliance will be determined and about the audit trail that will have to be maintained to be sure that people have made appropriate informed decisions where discretion is allowed.

I would take exception to a comment that has been made two or three times here. I do *not* believe there should be as much flexibility as possible. I believe there should be as much flexibility as is necessary to achieve two outcomes: first, to make sure that the arrangements that are made are the best context-specific arrangements required to promote welfare and good science and, second, to allow innovation so that we can continuously challenge and increase standards. But if you want to be an innovator, I think you also must be aware that the onus is on you to demonstrate the impact of your innovation, rather than simply to blaze a trail and hope others will follow.

We have also heard there are many things that can influence a data set, not all of which are recognized or captured at the moment. I will take you back again to the Russell and Burch book of 1959. They recognized the same thing. One of the things they pointed out about the Methods section of the publication is that your duty as an author is to use the Methods section not simply to describe what you did, but to provide enough detail so that other scientists can do what you did and see what you saw. You should provide ample information if you want others to be able to reproduce your work.

Let me leave you with this final thought, from someone who is a regulator. There are many gaps in the science behind the guidelines. In the real world, science, at best, informs rather than determines the outputs and the political process. But let me tell you this: In the absence of science and evidence, it is those with influence, not those with knowledge, who determine the output of that process.

DR. MILLER (John G. Miller, American Association for Accreditation of Laboratory Animal Care (AAALAC), International, Rockville, MD): As the Executive Director of AAALAC International, my position regarding harmonization is somewhat indescribable. Prior to this meeting, the discussions about harmonization of which I was aware focused on harmonization of standards. In that context, I think my "bottom line" is that there should be no mutually exclusive provisions, such as *never* allowing animals to be single housed.

We have already heard some of the other factors that must be considered in any discussion of harmonization, such as cultural and socioeconomic differences. It is difficult to convince individuals and organizations in developing countries that animals should live better than a large percentage of their population. This difficulty might lead one to the least-common-denominator approach, but is that really where we want to be? We just heard from Dr. Richmond that he does not endorse that approach, which leads me to "Miller's law." Miller's law states that the level of specificity in a standard, or indeed in any written document, is inversely proportional to the number of different viewpoints involved in the preparation. In other words, complete agreement results in lack of substance.

The exception to the Miller's law equation are the principles of the Committee of International Organizations of Medical Science (CIOMS) (CIOMS 1985). How many here know what the CIOMS principles are? As I suspected, very few of you are indicating that you know. They are excellent, and already agreed upon. CIOMS is part of the World Health Organization, so the United Nations and many countries are involved, and their principles are as specific as a multinational agreement can ever be. My position is that the likelihood of formulating meaningful harmonized written standards is extremely low, and even if it were possible in the real world, as Dr. Richmond points out, their application and implementation would undoubtedly be uneven, at best.

So what might be possible? I think it is possible to have harmonized *practices*. We already have examples of these practices—ethical review, animal care and use review, and national oversight authority—which can be harmonized and upheld by a recognized national authority. The practices can be based on harmonized principles; the CIOMS principles could be rejuvenated. An international organization such as AAALAC would be involved to assess and verify that the practices are, in fact, harmonized to a single standard that is not written, that is evolutionary, and that can change as we learn more about the area of animal care and use.

DR. GARNETT (Nelson L. Garnett, Office of Laboratory Animal Welfare (OLAW), National Institutes of Health (NIH), Bethesda, MD): It is extremely important to recognize that the mission of OLAW is animal welfare in the context of the mission of NIH, which is biomedical research.

Partly for that reason, I believe that harmonization is extremely important for both scientific and welfare reasons, if only, for example, to improve the quality, reduce the variability, and increase the ability to compare results across countries. With regard to welfare, in addition to the potential benefits to the animals themselves, we have the issue of credibility and the integrity of the process for which we are striving to gain public support. Properly applied, harmonization could enhance all of those various things.

I do think it is very important that we have some recognition across jurisdictions about equivalency of several systems or international standards. Actually NIH already does that. Although most of the NIH funding is within the United States, there are nondomestic institutions that receive NIH funding, either directly or as performance sites. The recognition of the sovereignty and, essentially, of the standards of the performance-site country is built into our system. That recognition is firm, and it requires people to follow their own rules.

I agree that it is important to avoid engineering-based mandates that have numerous implications, both intended and unintended. I am thinking in particular of things like trade barriers, wherein artificial restrictions or limitations on the use or the sale of a product, for example, might be limited. I believe we need to avoid such examples.

I agree that we need science-based standards to the extent possible, but it is still extremely important to do the science. As we have heard over the last 3 days, interpreting what the science actually means requires tremendous care. What is it telling us? Are we looking at the results objectively? Perhaps more importantly, are we conscious of the fact that some self-interest may be involved? We must be careful to separate ourselves from conflict of interest in how we interpret and apply the knowledge we gain. Finally, I believe it is extremely important that we as a community identify the questions that need to be answered to inform public policy in this area.

DR. DELEEUW (Wim deLeeuw, Directorate of Inspection, Food & Consumer Safety Authority, Department for Veterinary Public Health, Animal Diseases, Animal Welfare & Feed, Den Haag, The Netherlands): As a public health inspector in the Netherlands and Chairman of the working party of the Council of Europe, I also am in favor of harmonization. I will not repeat the arguments already mentioned; however, I suggest that we might question the kind of harmonization we are discussing. First, in my opinion, it is not necessary to strive toward a harmonization of *all* of the regulatory instruments because, indeed, cultural and international differences play an important role. An attempt to force all of these differences into one process would not be appropriate and would be a very long process. Instead, I believe that harmonization should target starting points—criteria and opinions. Especially in the preparatory phase of such a process, this approach will lead to the exchange of opinions and thoughts, as we have experienced these last few days. That process in itself will lead to improvements of quality. If the process goes well, then we can take the best from several worlds. Whether the laboratory is in Berlin or in Washington, I believe there is a common goal—to define and provide for the needs of the laboratory animals. To that end, our common goal requires common, shared knowledge.

For me, and perhaps for the majority here, the benefits of these meetings have been quite variable. We had a very good exchange of information about the regulatory processes in our countries and on our continents, about ongoing revision processes, and about results of relevant research. Yet at the same time, it has been evident at this meeting that many differences still exist with regard to opinions, knowledge, and perception. The question is how we can progress in striving for harmonization. One idea is to reduce the attendance at future meetings in an effort to create a kind of intercontinental or transatlantic steering committee. A smaller-scale group could be responsible for staying on course, for taking initiatives without having to do all of the work themselves, with the aim of making their goal more visible. In this meeting, I have perceived a significant lack of knowledge in some areas, whereas in other areas the amount of knowledge appeared to be great, but either it is not available everywhere or it is not accepted to the same degree everywhere. So perhaps a smaller group could take the initiative to strive toward reaching agreement on research criteria for best practices, which could form the basis of harmonization, at least in opinions.

DR. FORBES (Derek Forbes, Division of Biomedical Services, University of Leicester, Leicester, United Kingdom): We have heard much discussion this week about the gaps we need to fill in science-based data. I believe some of those gaps are created by a lack of documentation of the science that has already been done. With regard to safety testing, I would take the Organisation for Economic Co-operation and Development (OECD) as an example of international regulations that are based on good laboratory practice. The essence of good laboratory practice is its documentation, which can be questioned and inspected. Different countries reciprocate each other's inspectors. The quality of the safety testing is paramount because the products derived from those safety tests and efficacy tests are applied internationally.

In terms of harmonization, I suggest that those of us who use animals, at whatever level of research, should have the benefit of sufficiently well-documented studies to respond to any type of interrogation. The scientific quality of the studies should be paramount. In addition, the scientific journals should assist us by publishing sufficient information for the reader to be able to judge the quality of the science and interpret it. This type of cooperation among scientists would close many of the gaps we have discussed.

DR. GAUTHIER (Clément Gauthier, Canadian Council on Animal Care (CCAC), Ottawa, Canada): I would like to congratulate ILAR and the National Academy of Sciences for having held this conference, which has been extremely useful. On behalf of CCAC, I would like to express *qualified* support for harmonization. By harmonization, we really mean harmonization of national standards, of guidelines, and of a concerted approach to reach common objectives, rather than "international standardization." I agree with what Dr. Miller said about non-mutually exclusive positions. I believe that these guidelines should have the following three characteristics: They should (1) provide clear benefits to the animals, (2) not interfere with research, and (3) be based on science. But here I would like to depart slightly from my predecessors' statements. In addition to the publications written by authors who represent the conventional scientific disciplines (e.g., behavioral sciences, laboratory animal sciences) and all of the different types of knowledge that are not necessarily published in the mainstream scientific journals, I believe we should also recognize the side of scientific information that is coming and needed. Guidelines should be published and used as reference tools.

I recognize that some people might interpret the published guidelines as regulations, but I believe that this situation is a matter of perceptual distortion, which takes place at the individual level. I have faced that perceptual distortion in Canada several times, with individuals, for example, who are less qualified to be on an animal care committee than other members. Rather than understanding the basis and the rationale for CCAC guidelines, and because they often have not been involved early enough in the process, they try to place the onus on the CCAC. Regulatory agencies also face this situation, which I believe is a matter of perception. I agree with Dr. Richmond that in this context the people are that key, and their professional judgment makes the difference. This was the situation 30 years ago, and it is still the case, that we have insufficient collective knowledge to be able to draft, write, and agree on specific guidelines or on standards. We defer to professional judgment at the local level, and the substantive ethical decisions are made at the local level. Even if the scientific community is ready to progress collectively and worldwide, it cannot progress beyond the local level to another stage in the guidelines.

In conclusion, I would like to make the following three points: (1) We need to maintain a sensible basic approach throughout institutional animal care and use committees to develop and implement guidelines that incorporate the input of our peers. At the outset—not 10 years after the guidelines have been developed—we should include scientists, welfare organizations, and the public. (2) We need to exchange good practices across animal care committees and through international networks. Animal users in all countries should increase their understanding of their own national regulatory/oversight system. I have perceived a major deficiency in this context during these last 2 days of debate. If one lacks a basic understanding of how those guidelines are developed and when they should be applied, one cannot apply them or know their range of flexibility. Guidelines have flexibility in European countries. For example, in Europe I gained an understanding of the ban on the ascites method, which is equivalent to what we have in our recently developed guidelines. (3) We should increase the communication between regulatory administrators across countries on an ongoing basis. Dr. de Leeuw made this proposal, which Drs. Miller and Garnett also addressed. One simple way to accomplish that goal is to invite each other to hands-on experiences at the local level. For example, Dr. Garnett participated in one of our assessments at a large institution in Canada about 2 years ago, after which he basically confirmed that CCAC accreditation would serve the purposes of the Public Health Service.

DR. CRABBE (John C. Crabbe, Oregon Health and Science University and the Portland Veteran Affairs Medical Center, Portland, OR): I volunteered to speak against harmonization. Because you may characterize my position as essentially a straw-man argument, I will clarify it. I believe that the drive toward harmonization will inevitably lead to standardization. My concern is that standardization will further restrict the environmental range in which we study organisms that serve as our experimental subjects.

My particular interest is in understanding how genes influence behavioral endpoints, which leads inevitably to misunderstanding completely how genes, in fact, influence behavior, and under which circumstances. It can easily lead to false-positive findings, wherein one concludes that a gene affects a particular behavior. However, if everyone is using the same cage size and the same lack of enrichment, we are really studying the effects of those genes under a very small window of the total environment. It can also lead to false-negative findings, wherein one fails to detect true influences of either the biological manipulation in use or the genetic manipulation in use, because you looked in only one place, and the gene acts in many other places. Generally, because genes and biology interact with the environment of the animal, we need to explore environments and the biology of the organism systematically.

DR. KAGIYAMA (Naoko Kagiyama, Central Institute for Experimental Animals, Kawasaki, Japan): I would like to take a moderately against stance. I wonder to what extent we can harmonize the guidelines. Fortunately, the 3Rs principle has been adopted worldwide, so we may easily harmonize that principle in the guidelines. But what about the process of realizing the principle? Do you think we can harmonize the process to reach the goal? I daresay it is not easy. Each country has its own legal requirements, culture, and attitudes, which extend beyond science, are sometimes political, and are sometimes personal.

Actually, both the European countries and the United States are guiding the process independently by the ILAR *Guide for the Care and Use of Laboratory Animals (Guide)*. ILAR has chosen a science-based performance approach. In reference to the approach in the *Guide*, I would like to mention and paraphrase an interesting sentence that appears on page 25: It says that animal performance indexes, such as health, reproduction, growth, behavior, activity, and use of space, can be used to assess the adequacy of housing. The *Guide* thus advises us to observe animals precisely to know how to treat the animals. I understand this to be the basic concept of a performance approach, and I believe that we might harmonize a process if we focus on a performance approach. In that case, dissemination of the latest knowledge and technology, as well as the information about the variety of regulations, would be essential for the harmonized guidelines. For these reasons, I am moderately against harmonization.

DR. STOKES: Thank you, Dr. Kagiyama. At this point, I would like to return to Dr. Miller, who wanted to add a few clarifying remarks about AAALAC and to pose a few questions to the rest of the panel.

DR. MILLER: I am a firm believer in the old marketing adage, that "it is bad publicity only when they misspell your name." From that standpoint, I appreciate the frequent references to AAALAC during this workshop. Although I think that AAALAC's record, utility, and worth speak for themselves, I also believe there are some misconceptions, which I would like to correct for international colleagues who are less familiar with AAALAC that my US colleagues, and who may benefit for a brief review of AAALAC's role.

To begin with the basic system by which science is funded in the United States, people and companies pay taxes. The government decides how to allocate those funds, and it gives multiple billions of dollars every year to NIH, which then funds individual researchers and institutions. As in most instances, when you take public money, "strings" are almost always attached. In spite of the fact that we have a very decentralized system, some strings are attached. When you accept federal money to do research, you agree and promise, through your written assurance to Dr. Garnett's office, that you will follow the provisions of the *Guide*. You are not able to pick and choose which provisions you will follow and which ones you will ignore. You commit to follow them; therefore you must follow them. However, we have a system that specifically allows deviations from the provisions in the *Guide*, as long as they are scientifically justified and approved by the institutional animal care and use committee (IACUC). So there is a mechanism in place to deviate from what you have agreed to follow.

It is also very important for all of us to be reminded—because we tend to forget—that both the Public Health Service policy and the Animal Welfare Act put science above other requirements. Science will always prevail. The intent of both is to never interfere with science.

The role of AAALAC in this context is to assess the institution's adherence to agreed-upon standards. As mentioned, you have voluntarily invited visits from Dr. Garnett's office for their provision of an independent assessment. We therefore hold you to that standard, which has flexibility.

Finally, the *Guide* we use does not require transparent cages. It requires that cages allow observation of the animals with minimal disturbance to them. So if there is a way to accomplish that provision without cages being transparent, then you are free to act accordingly. I would encourage you to get your copy of the *Guide* and refer to page 36 to read about environmental enrichment. It is extremely flexible, as are most sections of the *Guide*, and it is open to a great deal of interpretation at your institution.

With regard to harmonization, I believe that many people are undecided at this point, and very few people enthusiastically support the concept. For those who are in favor of harmonization, I have the following series of queries, organized into three main questions: (1) If you accept my premise that there can be no mutually exclusive provisions in harmonized standards, then how can standards that contain different engineering standards (e.g., cage sizes) ever be harmonized? Can harmonization of the Council of Europe ever occur, based on the recently established cage size requirements? Compare those requirements with the provisions of the *Guide*. (2) Is it possible that you consider the general provisions of ETS 123 in the directive to be very similar to, even congruent with, the *Guide*, but that when you evaluate the specific wording, the two are different? If different engineering standards are permitted in so-called harmonized standards, may multinational companies (e.g., Dr. Kastello's company) allow the least stringent to meet the local requirements? Can they meet the requirements in Europe with the European requirements and in the United States with the US requirements? The preceding concern leads to the last question. (3) If the answer to question 2 is "yes, certainly they can," what are the implications for the business operation of those organizations? Will they move their animal studies to locations with the least stringent standards, or will they, in fact, contract the work out to other countries with even less stringent standards?

DR. RICHMOND: I stated my position earlier as "yes, but"; however, when you have been married as long as I have, you know that "yes, but" is very similar to "no, unless." To correct a misconception about the European situation, in fact, the Europeans will not be required to use particular cage sizes. In truth, the revised documentation will determine minimum sizes, but it will not dictate the size of the cage.

I think if you start from different positions, if there are economic as well as science and technical arguments to take into account, and if there is a lack of evidence on a number of the key provisions, it is necessary to work through a multi-staged process to achieve harmonization and to remain focused on the outputs. I think the first stage is to try to agree to the principles that should apply. I think everyone has agreed that good welfare and good science are worthy achievements. If we can agree to the principles, we might then be able to think about defining the processes that are needed to achieve those outputs—the mechanisms we might have to use, the judgments we might have to exercise, the evidence we might have to review. Those processes could be considered the second stage of harmonization.

If that level of harmonization achieves the desired outputs, we could stop. If after achieving that level of harmonization, no one has changed and everyone behaves as before ("because it has always been done this way"), I believe the regulators will decide to harmonize the practices. If the situation progresses to that stage, it will be the scientific community that has failed to adjust after the principles and the processes were harmonized. If it is possible to achieve the outputs through principles and processes, regulators have no reason to prescribe the practices.

DR. DELEEUW: I believe that discussions about harmonization of specific, quantitative matters such as cage sizes should be the last step of the process. In general terms, as Dr. Richmond said, we agree on the starting points, the principles. Where we might not always agree is on the steps between. We saw such examples in the descriptions of the Council of Europe proposals for rodents and rabbits, in the description of how much space is needed for certain types of enrichment. That example is the minimum freedom of movement that an animal should have. However, if we are able in the future to agree on basic principles, then the last step, harmonization of cage sizes, should not be a real problem. I believe it depends on the steps between, the translation of the principles to behavior.

DR. STOKES: I agree, and I think you have outlined a good process. However, it appears that we have moved in that direction without an inclusive process. Based on that premise, I would like to ask Drs. Richmond and de Leeuw the following question: Within Europe, with passage and adoption of these Council of Europe guidelines, is there the opportunity for exception, as Dr. Miller mentioned, to deviate and, in some circumstances, perhaps even use a smaller cage because there is a scientific rationale (e.g., to replicate data for a similar study that was conducted in the United States, with smaller cages)?

DR. RICHMOND: I think the answer is yes, that there will be the scope to do that. However, I am not certain of the actual specific arguments. The specific case you have described will not necessarily be accepted.

DR. DEMERS: When we are talking about cage size, we are talking more about standardization than about harmonization. I believe that we can agree on some specific concepts and some ideas. Some things depend on the culture (e.g., small hotel rooms in Japan vs. large hotel rooms in the United States—basically minor matters), and other things depend on economic purposes (e.g., exchange and trade); however, I do not know whether other things have an effect on science.

DR. STOKES: Dr. Gauthier recommended a process that would allow input from all of the involved stakeholders—the public, animal welfare groups, scientists, animal care specialists. Would any of the panelists like to propose how that process might be accomplished?

DR. GAUTHIER: I made that proposal because we already function this way in Canada. The process of guideline development involves first, of course, the scientific community, when we establish a subcommittee to develop a specific guideline; but also at the time of the second draft, we open it to public input. At that time, we also open it for other countries to have input. Most of the time, however, we ensure harmonization by including members of other countries in the subcommittee. This is the case, for example, with the subcommittee for fish and for other guidelines, when we are across borders with our southern neighbors and are taking action on the use of animals, where we have joint interface.

Also in our case, guidelines do not become standards per se until we have tested their rigor and solidity at the assessment level. On our peer-based team of assessors, we include representatives from the public who have been nominated by the Canadian Federation of Humane Societies, which entails additional input. We also have representatives on the institutional animal care committee, ranging from animal technicians to students. They all bring their input into the decision-making process and into these feedback learning loops that I mentioned. So it can be done. As we assess institutions and the basis of the guidelines, we find weaknesses that have been identified by peer-based assessment teams, which we later correct. We incorporate those items 2 years later as part of the review of the specific guideline.

This feedback mechanism is extremely useful. By the time we have completed the revision, we have already incorporated what the scientific community told us through the tour of our 190 participants in the program. There is no communication break per se; however, that continuity is actually part of the problem. The same thing is happening between countries, which is why I was calling for extended communication during guideline development with other countries, where it is relevant.

This mechanism should begin at some point. Then you see the value of it as you go through it. In our case, what it did for us was that the acceptance of the guidelines developed by everyone has reached a level now where government regulators are using the guidelines in areas where they have no specific jurisdiction to take action, or the public puts pressure on them to do it. We have extended the loop to government regulators, to the point where everyone is now included, including the regulators. So the process does work, with guidelines and standards.

DR. MILLER: Before we bring together a group of stakeholders, I believe we need to step back and ask why it is important. Why do we want to harmonize whatever we are harmonizing—guidelines or processes or whatever? What is the negative impact, and on whom is that impact negative, of not having harmonized guidance in this area? In other words, who cares? We could have a meeting like this, and we could all come together and agree on steps in harmonizing a process, but who would benefit? I think we have to answer those questions first, and know that there is a driving force for the effort. Perhaps Dr. Kastello would like to comment because he probably represents an affected group.

DR. KASTELLO: Certainly we would be affected by standards that would cause significant economic impact on the way we presently operate. Those of you who know me know that I am committed to animal welfare and to doing the right thing. As long as I have that responsibility in my organization, we will act accordingly. But I have to face the realistic problems of the economics. If the Council of Europe space revisions become part of the directive, they will significantly adversely affect my ability to house the numbers of animals I need to house for our present volume of studies in Europe. What does that mean to me? What are my choices? I can buy larger enclosures, I can build more facilities, or I can move the science somewhere else, where it is more cost-effective. Obviously, I will not be making those decisions, but we are talking about a great deal of money for one company. I speak only as a representative of a pharmaceutical company, but the entire industry will be faced with these kinds of decisions.

DR. GAUTHIER: I am reminded of what was said by one rapporteur this morning, which I believe is key. In assessing the value of any guideline or standard, rather than a single element like the measure of a cage, we should consider a program of environmental enrichment for the animals. The program should include, but not be limited to, cage size. This area is where we fail. We have a tendency to have a stovepipe approach, to limit what we think into one area. It is very scientific to try to exclude all other factors and retain only one factor, but the one we keep might be the one with the lesser significance. We should return to the concept that we are looking for and assessing a program, not simply the single element of a program.

DR. STOKES: At this point, I would ask for any comments or questions for the panel from the audience.

DR. ELLIOTT (Rosemary Elliott, Roswell Park Cancer Institute, Buffalo, NY): I would like to comment from the perspective of a mouse user. During this meeting, Dr. Smith presented data suggesting quite strongly that the current number of mice one is allowed to keep in a shoebox cage is very conservative. She estimated a number that was more than twice as large, which resulted in no major problems with the mice she was using, although admittedly the mice were a limited age range and a single strain. However, if these kinds of data are validated in the future, how will the various organizations respond? Can we put more than five mice in our standard cage? How will the European group respond? These are data. How are you going to work with these new data that will be coming out?

DR. GARNETT: I believe these data will clearly influence the existing processes, for example, what we hope will be a revision of the *Guide* in the next few years. The *Guide* revision committee should be looking at valid peer-reviewed science in formulating its collective judgment as to what is appropriate. So the more science, the better. I think many of these cage sizes we are dealing with now may not be supported by the hard science to which you are referring, although there was, obviously, consensus at some point in time. It is part of the existing process and should inform the decisions.

DR. MILLER: AAALAC has traditionally been a kind of "show me the data" organization. I believe there are some current, and I know some past, ILAR Council members here who might also address this issue. Dr. Golding, in fact, mentioned in one of the breakout sessions yesterday that AAALAC had identified what they considered a problem with three mothers. It was not a mandatory item, because we did not have the data. They kept three mothers and litters in one cage, and because it appeared that they were crowded, the suggestion was made to investigate, which they did. The resulting data indicated that the numbers were, in fact, detrimental; and they therefore discontinued that practice. Importantly, we are always seeking data. Dr. Smith's data—in her hands, in that strain, and under those conditions—were very compelling. I think it is relatively safe to say (although I never can speak for the Council on Accreditation) that if you were to reproduce for them Dr. Smith's data, they would find the data acceptable with respect to that density of animals.

DR. WITT (Clara Witt, Department of Defense, Global Emerging Infections, Silver Spring, MD): Based on my previous laboratory animal experience, I believe that Dr. Miller has brought up a very good point about the CIOMS. I have been a consulting veterinarian for the International Agency for Research on Cancer in Lyon, France—a United Nations (UN) organization animal facility that is supranational and not subject to the rules or regulations of any national jurisdiction. I too ask the following questions: What do we do with an animal colony? How do we apply for an NIH assurance letter for things like this? It was also necessary to work with the practicalities of everyday life—funding, good care of animals, what we do with our researchers, how we work with our researchers.

I agree with Dr. Miller that it may behoove us to look very carefully at the CIOMS principles, which served the UN animal facility very well. They provided the principles, the starting block for an international body of researchers, animal caretakers, and administrative staff, to work toward a continually improving animal care and use program. It was not perfect, and our animal care and use committees were "interesting," to say the least, because various cultural, previous-research-background perspectives were involved. Yet the principles were solid, and we were able to work out many issues and eventually agree on some degree of standardization. For this reason, I do not use harmonization in the same sentence with standardization, because I think it blocks our ability to achieve harmonization toward the basic principle of good science and good animal welfare. I believe we all need to look at those principles very closely. It will be fantastic if at some point it is decided within the communities that those principles need to be updated and refined, although at this point they do provide a very good starting point for animal care and good research. [NOTE: See Appendix D.]

DR. MILLER: I am not certain whether the principles are online, but I propose including them on the ILAR web site. There are other multinationally agreed-upon guidance points, such as those of the International Conference on Harmonization, which involved Europe, Japan, and the United States, although those points are essentially restricted to pharmaceuticals. Dr. Richmond and I also worked together on an International Organization for Standardization (ISO) technical committee that has just completed (with Dr. Richmond's excellent leadership) a revised set of animal welfare standards: ISO standard 10 993, Part 2, Animal Welfare. There are approximately 80 members of that committee, and many countries have agreed to the very basic tenets. So if, in fact, it is a good thing to harmonize processes, we will at least have many resources and examples. The good laboratory practices of OECD, with which many people agree, also relate to animal welfare.

PARTICIPANT: As information, I would like to add that the CIOMS principles were developed under the auspices of CIOMS as part of the World Health Organization, and were chaired by a friend of all of us, Dr. Held. At that time, Dr. Held was chairing an NIH interagency committee, which redrafted the CIOMS principles into the US Government Principles, which you will find in the back of the *Guide*. These principles are the tenet of the Public Health Service policy.

DR. GARNETT: In addition, most of us in this room belong to professional societies, which also provide statements of principle or policy, or editorial review principles that paraphrase or restate almost verbatim the CIOMS and the US Government Principles. So they are already deeply entrenched in our minds.

DR. STOKES: I would like to add that those principles clearly incorporate Russell and Burch's principles of refinement, reduction, and replacement—the 3 Rs. These principles are clearly communicated.

DR. STEPHENS (Martin Stephens, Humane Society of the United States, Washington, DC): Based on the premise that animal research is a societal activity (i.e., currently more or less sanctioned by society, and regulated by governments that are more or less accountable to the people, at least in principle), then it follows that the stakeholders should be involved in whatever guidelines are written, whether it is within or across countries. Many of you have mentioned that animal welfare groups should be part of the process, and I believe their inclusion gives the process a certain legitimacy, because the result are guidelines that are carried out within the walls of research institutions, where it is difficult for the public to know what is happening. However, if animal welfare representatives are involved at the front end, there will be more assurance that the guidelines themselves are at a decent level.

We would like to see guidelines that have enough substance and specificity that there is some guarantee of a minimal standard. I do not personally like the minimal standard approach, but the public should be given some sort of guarantee that facilities cannot interpret those guidelines and basically do nothing. We know that some facilities are motivated to do the right thing and to do good, but we are particularly concerned about other facilities, which may have other priorities and allow standards to "fall through the cracks." In addition, we are concerned not only about the drafting but also about the application of these guidelines. They should have enough substance so that facilities that basically flout them face some consequences, and so that the public has some guarantee that the system will ensure accountability.

DR. GAUTHIER: Dr. Stephens has stated a very important point regarding both the substance and the implementation of guidelines. I also mentioned this point in my presentation, when I talked about the Canadian mice that are more stressed than the American mice. I mentioned the recognition, during the June 2001 International Symposium on Regulatory Testing and Animal Welfare, that the ACC system is a very good system for implementing guidelines locally. Based on the work of the committee that Dr. Richmond chaired, most of the 22 countries now have reliable implementation systems in place that are being legislated by or based on a voluntary framework. We are not able to conceive and implement some harmonization measures, because we can trust that in all of those countries involved with harmonization, there is a system in which a third party can cross-check the implementation of the guidelines.

DR. BLOM (Harry Blom, Utrecht University, The Netherlands, and Vice President of International Liaisons, Federation of European Laboratory Animal Associations (FELASA)): I would like to mention an issue that, to my knowledge, has not been addressed in this meeting but that can have profound effects on the outcome of experimental results: how people have acquired their expertise and skills. As we all know, countries have different requirements for training and education, and when people move from one country to another, even though they have been trained, they often must complete additional training to practice in the new country. I would like to discuss this issue, and consider possible harmonization of education and training standards. I would like to hear from members of the panel on the issue.

DR. RICHMOND: Dr. Blom was too modest to declare a potential conflict of interest. Within Europe, FELASA has produced guidelines for key people within the animal care and use sector. Although the European institutions have not endorsed them, individual European countries have in most cases endorsed them. Within the United Kingdom, for example, we have our own training requirements, which are essentially theoretical. We do not allow practice on live animals (with one exception, which I will not detail at this time), and we require both basic skills and proper supervision and training in bona fide testing and research programs. We are uneasy with people having boxes of practice animals for practice in facilities.

We recognize relevant training from any other European countries, provided there is documentary evidence, and we have one additional requirement—completion of module 1, which is essentially an overview of UK domestic legislation. We do not expect a visitor from overseas to be equipped with that information in advance. We have recently accredited an Internet distance-learning course that will allow people to undertake the training before they arrive.

DR. MILLER: Dr. Blom's suggestion is excellent. As you know, AAALAC already has included the FELASA guidelines in our reference resources. In the United States, the IACUC must ascertain and confirm that everyone who is doing something with animals is appropriately trained and qualified. It is a very broad performance standard.

DR. GAUTHIER: In Canada, we have been promoting the training programs given by institutions. In 1999, we developed and published a guideline on the training of animal users, which became mandatory for assessment purposes in January 2003. We are now verifying, through our regular 3-year cycle of assessments of all institutions, that institutions do have the training structure in place. In addition, we provided assistance by publishing on our web site 12 modules on training for animal users— mainly researchers, graduate students, and research staff. The modules are a very good continuing education tool, which is available in both French and English. It is a mandatory requirement, but hands-on training, of course, is the responsibility of the institutions as per our guidelines.

DR. NELSON (Randall J. Nelson, Department of Anatomy and Neurobiology, IACUC Chair, University of Tennessee, Memphis): It is my responsibility to share these discussions with fellow IACUC members, and I will say that we agree on many issues. However, I submit that we still may need to deal with another aspect of implementation philosophy. I would describe the two sides of the issue as centralized versus decentralized views. To use the analogy of our own conflict about 140 years ago in the United States, these views might be analogous to states' rights versus federalism. I have heard that some implementations are very centralized, having a set of rules, guidelines, or prescriptions that must be accomplished by all involved, and they are being implemented at a national or supranational level. I have also heard of an implementation structure that seems to be diffuse or distributed to individual IACUCs and/or institutions, allowing those bodies to set certain internal standards, provided they do not deviate from the *Guide*.

In both cases, I can see benefits, for example, consistency in the centralized system versus markedly different implementations, presumably, at different institutions. I would like the panel to discuss the benefits and the drawbacks of these two systems. I believe it will be necessary to deal with those differences in philosophy before we can decide whether we can harmonize or standardize.

DR. MILLER: Dr. Nelson makes an excellent point and has identified the advantages to both of those types of systems. As I mentioned earlier, I view AAALAC as performing de facto harmonization, or implementation. I authored an article in *Lab Animal* in the late 1990s titled "Harmonization: The Proof Is in the Practice." In retrospect, the title should have been "The Proof Is in the Performance." You can have all of the desired harmonized standards and all of the different desired standards, but these desires are meaningless unless you monitor implementation. You have no system unless you have a way to verify that an institution has fulfilled its national requirements.

I take pride in AAALAC because in more and more countries, we assess performance based on a host of standards that begin with the respective national requirements. The basic principles of the *Guide* must always be met. You can be assured that our evaluation of performance under those circumstances is consistent to a great extent, that what takes place in an institution we accredit in India is very similar to what takes place in an institution we accredit in Indiana. So I believe it is possible to establish consistency. There are advantages and disadvantages to both, but if you want to harmonize process, you need some sort of national system. You do not need regulations, as Canada has shown us.

For those countries that do not already have a system in place, the Canadian system is absolutely the best, bar none. You start with some sort of national system. Then you establish a responsible authority, which includes a process for reviewing the ethics and the procedures of each proposal. The authorities can be in the same body or they can be separate bodies; they can be national, regional, local, or whatever—as long as they exist. Finally, you need a monitoring mechanism to confirm that requirements are being met. You can have all of the rest, but it will be a hollow exercise unless you have that last mechanism—whether, as in the United States, it is the Department of Agriculture coming in and doing unannounced inspections, or a voluntary program like AAALAC.

DR. GAUTHIER: The Canadian Council on Animal Care operates a decentralized quality control peer-based system through the institutional animal care and use committee. This system assesses the functioning of all of those committees as well as the facilities and so forth. The same people who are part of their own institutional animal care committee, at a different time, become assessors on one of our teams. They contribute by generating and circulating the knowledge from local to national, and from national to local levels. It is, again, decentralized quality control, or centralized quality assurance.

DR. RICHMOND: I believe that some things are best done, perhaps should only be done, at the top—things like policy and principles probably must be top-down, rather than bottom-up. However, processes and practices must be bottom-up. I cannot sit in an office 380 miles away from a research laboratory and know what is best for them to do in the context of what they are doing that day.

With regard to verification, I would like to repeat three key points: (1) If you are spending government money, and if you are at the user end, you must be absolutely clear about what you are actually being asked to do or what you have actually contracted to do. (2) You must have an intelligent regulator who has the degree of technical expertise to argue the technical points with you, rather than simply restate the regulation. (3) You need to add to your oversight system a mechanism that draws attention to things that are now going out of date and need to be replaced, and something that highlights the area where people have excelled and produced something that may be better than the common current standard.

PARTICIPANT: I would like the panel to suggest ways to close the knowledge gap, to generate the necessary data that we have discussed during this workshop. I would suggest that we consider and possibly adapt for our needs an existing model operated by the organization called International Life Sciences Institute (ILSI), which addresses various public health questions. When they have a question, they convene a meeting of the stakeholders who are interested in that question who then form a consortium. They look at the work that needs to be done, they divide it among themselves, and they then perform different experiments. The protocols are reviewed by a central steering committee, and the results are shared with the organization and eventually published. That system is one possible model for the laboratory animal community to use in addressing questions on animal welfare. We have many different strains and species, and more unknowns than knowns. One possible way of attacking the problem, rather than simply scheduling a meeting 5 years from now and admitting that there is still a dearth of knowledge, is for one of the nonprofit international organizations to step forward and entreat people to participate in a consortium, divide up the work, and start doing it. Stakeholders not only could do some work on their own, but a body could apply for public funding by forming such a consortium. Universities and other institutions that are not profit-making institutions could then participate in the process as well.

DR. KASTELLO: Some organizations provide funding in this area, for example, we heard from Ms. Cathy Liss that the Animal Welfare Institute is providing some funding to work in this area. Since 1996, the American College of Laboratory Animal Medicine Foundation has provided small research grants to work at filling these gaps. It is not a huge amount of money, but it is at least an effort to begin to understand more about the animals we are using. Clearly, that process needs to be improved, and it needs to be advertised more widely so that people know there are some funds available, and to encourage people to submit research proposals.

DR. MILLER: What is the source of the funding in the ILSI program?

PARTICIPANT: The stakeholders provide the funds. The companies themselves pay for the studies in which they are most interested. The studies are prioritized, the work is divided, and the companies participate in doing that work.

DR. MILLER: It sounds like an excellent model. We are always left with the question, where will the money come from?

DR. KASTELLO: The European College of Laboratory Animal Medicine has also started a foundation. FELASA has helped fund it. The intent is to fund similar research grants, to enhance the knowledge about the animals we use.

DR. GARNETT: I would like to share with you briefly my meeting with NIH Director Dr. Zerhouni, when he first went to NIH. In response to his request for my (and other office directors') highest priority, I described the need for funding to answer scientific questions to inform public policy in the animal welfare area. He was very interested and asked for examples. Among the examples I cited of questions for which we have insufficient data—questions for which we need scientific answers—was whether decapitation is humane. I also cited issues related to the appropriate uses of CO_2. I believe he took my responses very seriously, but he also needs to hear from people other than his own staff. I believe animal welfare is a worthy thing to communicate, and you may agree that this issue should be a priority for NIH to address.

PARTICIPANT: I think all of the funding sources mentioned are very helpful, but the problem I see is a lack of coordination. If you had a consortium that listed the work to be done, prioritized it, and then used all available funding sources, you could get much of the work done in 3 or 4 years.

DR. STOKES: In closing, I would like to thank all of the panelists for their participation and for preparing remarks on their initial positions regarding harmonization. Clearly, our discussions reflect that we share common ground on one aspect of harmonization—the international principles on the humane care and use of laboratory animals. It appears that we have some work to do regarding the harmonization of processes, standards, and practices, but I think that opportunity is in the future.

DR. KLEIN (Hilton J. Klein, Merck Research Laboratories, West Point, PA): I would like to echo Dr. Stokes in thanking all of the participants. The dynamic exchange that I have perceived over the last 3 days has been incredible. In case I am an example of anyone else in this room, I will mention that I came here with some trepidation and anxiety, and I have heard many discussions and comments about people being pleasantly surprised. I think one of the results of the workshop was to open key channels of communication, and we have learned that some of the perceived gaps are not as real as we had thought.

In addition to thanking the participants, I sincerely thank all of the speakers, and particularly our plenary speaker Dr. Crabbe, our banquet speaker Dr. Holekamp, every other speaker on the program, and especially members of our panel, who did a marvelous job. I think the sharing of knowledge was facilitated by the fact that we planned the workshop to be interactive and participatory, and I believe the breakout sessions enabled us to achieve that goal. In summary, from the perspective of ILAR, this meeting has been one of discovery. All of it has been about science. Science is defined simply as knowledge. The English word is derived from Latin, which denotes the systematic observation of natural phenomena for the purpose of discovering laws governing those phenomena. This workshop certainly was a phenomenon.

Some issues that arose were surprises, for example, the fact that we do not know our own regulations and our own country's standards as well as we should. We need to improve in that area, and I think this workshop is a step in that direction. We heard that message repeatedly. I would also like to emphasize that we all need to communicate these issues to the scientific community. We need them on our side to address the issues. I think we all have the combined common goal of improving welfare and performing good science, but we need the scientists in the room. We have had some discussions about how to accomplish that goal. The first thing to do when you return to your laboratories is to share this information with your colleagues. This issue is very important both today and for the future, because if we do not resolve the problems ourselves, the public will expect others to do it for us. I do not think any of us want to be in that position, because the privilege of using animals in research, worldwide, is at stake. To maintain that privilege, we must address the issues around it. Our understanding of the guidelines, what drives them, what creates them, the science behind them, and where we get the funding—all of the issues we have discussed—will preserve and protect that privilege and the stewardship that is part of using animals in research. On behalf of the ILAR Council International Committee, I again thank every one of you for making this a wonderful experience.

DR. DEMERS: I have been asked by my colleagues at the international level to express our many thanks to ILAR and to the National Academy of Sciences, to Dr. Hilton Klein, Chair of the Program Committee, and to all members of the Program Committee. We also thank Ms. Kathleen Beil and all of the support staff who have assisted us in the planning of this meeting. We thank those who were involved in the organization of the meeting, and especially ILAR Director Dr. Joanne Zurlo. Thank you for providing us with the opportunity to explain and describe how we see things. We all are seeking the same high-quality result regarding the welfare of animals, and we all are scientists. Finally, but very importantly, we thank the sponsor and cosponsor of this workshop, without whose support its success would not have been possible.

I hope that this initiative will be repeated. I urge all of you to attend the next FELASA meeting, to be held next year in Nantes, France, where the theme of the meeting will be internationalization and harmonization in laboratory animal care and use issues. This meeting will provide an opportunity to maintain the ongoing dialogue, because even if we agree or disagree on harmonization, I believe that most of us agree that it is important to keep communication active and proactive.

Appendixes

Appendix A:
International Workshop on
Development of Science-based
Guidelines for Laboratory Animal Care

FINAL PROGRAM

DAY 1

8:00-8:30 am	**Registration and Breakfast—** **West Lobby** **Sessions in Salon ABG**	
8:30-8:45 am	Welcome and Introduction	Peter A. Ward Hilton J. Klein
8:45-9:30 am	Plenary Lecture— Genes, Environments, and Mouse Behavior	John C. Crabbe
Session 1	Current Status— Identifying the Issues	Coenraad F.M. Hendricksen (Chair)
9:30-10:00 am	AAALAC International Perspective	John G. Miller
10:00-10:30 am	The Council of Europe	Wim deLeeuw

10:30-10:45 am **BREAK**

10:45-11:15 am The Role of the International   Gilles Demers
                Council for Laboratory Animal
                Science at the International Level

11:15 am-   Process for Change—
12:15 pm   Regulatory/Oversight Comparisons

                US—The PHS Policy on   Nelson L. Garnett
                Humane Care and Use of
                Laboratory Animals
                Rulemaking by APHIS   Chester A. Gipson
                Europe—Process for Change—   Jonathan Richmond
                Regulatory/Oversight
                Comparisons—Europe
                Japan—Japanese Regulations on   Naoka Kagiyama
                Animal Experiments—Current
                Status and Perspectives
                Canada—Process for Change—   Clément Gauthier
                Regulatory/Oversight in Canada

12:15-12:30 pm Building Credible Science from   Paul Gilman
                Quality, Animal-Based Information

12:30-1:30 pm   **LUNCH—Faculty Club**

Session 2   Assessment of Animal Housing   John G.
                Needs in the Research Setting—   Vandenbergh
                Peer-reviewed Literature Approach   (Chair)

1:30-3:30 pm   Introduction: John Vandenbergh
                Speakers:
                Frederick S. vom Saal: Phenotype in Mice:
                Effects of Chemicals in Cages,
                Water and Feed
                Graham Moore: Assessment of
                Animal Housing Needs in the
                Research Setting—Dogs/Cats
                Melinda Novak: Housing for
                Captive Nonhuman Primates:
                The Balancing Act

Markus Stauffacher: How to Identify the Laboratory Rabbits' Needs, and How to Meet Them in the Research Setting

3:30-3:45 pm	**BREAK**
3:45-5:00 pm	Discussion

DAY 2

8:00-8:30	**Breakfast—West Lobby**

Session 3	**Approaches to Current Guidelines—US and Europe— Salon ABG**	Randall J. Nelson (Chair)
8:30-9:45 am	The "Guide" Process: Approaches to Setting Standards	William J. White
	The Revision of Appendix A of the European Convention for the Protection of Vertebrate Animals Used for Experiments and Other Scientific Purposes: The Participants, The Process and the Outcome	Derek Forbes

10:00-11:30 am Breakout Groups (Chairs)—
Mice/rats
    (Axel Kornerup-Hansen)   Salon ABG
    (Rosemary Elliot—R)
Dogs/cats   Salon D
    (Robert Hubrecht)
    (Thomas Wolfle—R)
Nonhuman primates   Salon E
    (David Whittaker)
    (Randall J. Nelson—R)
Rabbits   Exec. Conf. Rm.
    (Vera Baumans)
    (Jennifer Obernier—R)
(R = Rapporteur)

11:45 am- 12:30 pm	Reports from Breakout Sessions	Salons ABG
12:30-1:30 pm	**LUNCH—Faculty Club**	
Session 4	**Environmental Control for Animal Housing Salon ABG**	William Morton (Chair)
1:30-2:30 pm	Rationale for current guidelines and basis for proposed changes	Bernard Blazewicz Dan Frasier
	European Guidelines for Environmental Control in Laboratory Animal Facilities	Harry Blom

2:45-4:15 pm	Breakout Groups (Chairs)—
	Lighting
	  (Harry K. Blom)
	  (Michael J. M. Stoskopf—R)
		Exec. Conf. Rm.

	Noise/vibration impacts
	  (Sherri Motzel)
	  (Hilton J. Klein—R)
		Salon D

	Metabolism/immunology impact
	  (Jann Hau and Randall J. Nelson)
	  (Stephen W. Barthold—R)
		Salon ABG

	Engineering issues/security
	concerns
	  (Bernard Blazewicz and
	  Dan Frasier)
	  (Janet Gonder—R)
	(R = Rapporteur)
		Salon E

4:30-5:15 pm	Reports from Breakout Sessions	Salons ABG
5:30-9:00 pm	**RECEPTION—South Gallery BANQUET—Salon AG** *Speaker* A View from the Field: What the Lives of Wild Animals Can Teach Us About Care of Laboratory Animals	Kay E. Holekamp

DAY 3

Session 5	**Environmental Enrichment Issues Salon ABG**	Emilie Rissman (Chair)
8:00-9:30 am	The Science of Environment Enrichment for Laboratory Animals	William T. Greenough
	Search for Optimal Enrichment	Timo Nevalainen
9:45-11:30 am	Breakout Groups (Chairs)— Mice/rats and rabbits (John Vandenbergh and Vera Baumans) (Jennifer Obernier and Stephen Barthold—R)	Salon ABG
	Dogs/cats (Graham Moore) (Janet Gonder—R)	Salon D
	Nonhuman primates (Carolyn Crockett) (Randall J. Nelson—R) (R = Rapporteur)	Salon E
11:30 am-12:15 pm	Reports from Breakout Sessions	Salon ABG
12:30-1:30 pm	**LUNCH—Faculty Club**	
1:30-3:00 pm	**Point/Counterpoint**—The Cases For and Against Harmonization to Balance the Needs for Conducting Research and Meeting Animal Welfare Concerns— **Salon ABG** **Chair: William S. Stokes** *Participants:* Michael D. Kastello, Gilles Demers, Jonathan Richmond, John G. Miller, Nelson Garnett, Wim deLeeuw, Derek Forbes, Clément Gauthier, John C. Crabbe, Naoko Kagiyama	
3:00 pm	Workshop adjournment	

Appendix B:
Workshop Faculty

Vera Baumans, DVM, PhD, Dipl. ECLAM
Professor
Utrecht University, The Netherlands, and
Karolinska Institute, Sweden

Bernard Blazewicz, PE
HVAC Engineering Manager
Merck Manufacturing
USA

Harry J. M. Blom, PhD
Laboratory Animal Science Specialist
Utrecht University
The Netherlands

John C. Crabbe, PhD
Professor, Behavioral Neuroscience
Oregon Health and Science University
USA

Carolyn Crockett, PhD
Coodinator, Psychological Well-Being Program
Washington National Primate Research Center
University of Washington
USA

Willem deLeeuw, DVM
Senior Veterinary Public Health Officer
Ministry of Public Health
The Netherlands

Gilles Demers, DMV, MSc
President
International Council on Laboratory Animal Science
Canada

Derek Forbes, DVM&S, MSc, PhD, DLAS, MRCVS
President
Federation of European Laboratory Animal Science Associations
United Kingdom

Dan Frasier, PE
Director of Commissioning Services
Cornerstone Commissioning, Inc.
USA

Nelson L. Garnett, DVM, Dipl. ACLAM
Director, Office of Laboratory Animal Welfare
National Institutes of Health
USA

Clément Gauthier, PhD
Executive Director
Canadian Council on Animal Care
Canada

Paul Gilman, PhD
Assistant Administrator for R&D
U.S. Environmental Protection Agency
USA

Chester A. Gipson, DVM
Deputy Administrator for Animal Care
Animal and Plant Health Inspection Service
U.S. Department of Agriculture
USA

William T. Greenough, PhD
Professor of Psychology, Psychiatry, and Cell and Structural Biology
University of Illinois
USA

Axel Kornerup-Hansen, DVM, Dipl. ECLAM
Professor, Division of Laboratory Animal Science & Welfare,
    Department Pharmacology & Pathobiology
Royal Veterinary and Agricultural University
Denmark

Jann Hau, MD
Professor, Division of Comparative Medicine
Uppsala University
Sweden

Kay E. Holekamp, PhD
Professor of Zoology
Michigan State University
USA

Robert Hubrecht, PhD
Deputy Director
Universities Federation for Animal Welfare
United Kingdom

Naoko Kagiyama, DVM, PhD, Dipl. JCLAM
Expert, Animal Welfare Compliance
Central Institute for Experimental Animals
Japan

Michael D. Kastello, DVM, PhD
Vice President
Global Laboratory Animal Science and Welfare
Aventis Pharmaceutical
USA

John G. Miller, DVM
Executive Director
AAALAC International
USA

Graham Moore, DVM&S
FELASA Coordinator of CoE Expert Group on Dogs and Cats
Pfizer Global R&D
United Kingdom

Sherri L. Motzel, DVM, PhD
Director, Laboratory Animal Resources
Merck Research Laboratories
USA

Randy Nelson, PhD
Professor of Psychology and Neuroscience
Ohio State University
USA

Timo Nevalainen, DVM, PhD, Dipl. ECLAM
Professor
National Laboratory Animal Center
University of Kuopio
Finland

Melinda Novak, PhD
Chair, Department of Psychology
University of Massachusetts
USA

Jonathan Richmond, BSc, MB ChB, FRCSEd
Chief Inspector
Home Office
United Kingdom

Markus Stauffacher, PhD
Head of Ethology, Animal Husbandry and Animal Welfare Unit
Institute for Animal Sciences
Swiss Federal Institute of Technology
Switzerland

John G. Vandenbergh, PhD
Professor, Emeritus
North Carolina State University
USA

Frederick S. vom Saal, PhD
Professor of Biology
University of Missouri
USA

William J. White, VMD, Dipl. ACLAM
Corporate Vice President
Veterinary and Professional Services
Charles River Laboratories
USA

David Whittaker, BVM&S, DLAS, MRCVS
Director, Laboratory Animal Sciences
Huntingdon Life Sciences
United Kingdom

ADDITIONAL AUTHORS

Ann Benefiel
University of Illinois
USA

Sarah L. Mabus
The Jackson Laboratories
USA

Cameron Muir
Department of Psychology
Brock University
Canada

Susan C. Nagel, PhD
Department of Biological Sciences
University of Missouri

Tatsuji Nomura, MD
Director
Central Institute for Experimental Animals
Japan

Jan Lund Ottesen, DVM, PhD
Novo Nordisk
Denmark

Catherine A. Richter, PhD
Department of Biological Sciences
University of Missouri
USA

Rachel R. Ruhlen, PhD
Department of Biological Sciences
University of Missouri
USA

Abigail L. Smith, PhD
The Jackson Laboratories
(present address) University of Pennsylvania
USA

Jason D. Stockwell
The Jackson Laboratories
USA

Wade V. Welshons, PhD
Department of Biological Sciences
University of Missouri
USA

Appendix C: Committee Bios

PROGRAM COMMITTEE

Hilton J. Klein, VMD, MS, Dipl. ACLAM, Dipl. ECLAM (*Chair*), is Senior Director for Comparative Medicine, Merck Research Laboratories, and Adjunct Assistant Professor, Department of Laboratory Animal Resources, University of Pennsylvania. He has an extensive background in laboratory animal medicine. His research interests are in laboratory animal science, particularly in the field of laboratory animal infectious disease and surgical production of animal models. He has been a consultant to the Pan American Health Organization as Merck's representative on nonhuman primate conservation.

Stephen W. Barthold, DVM, PhD, is Director, Center for Comparative Medicine, UC Davis, Director, UC Davis Mouse Biology Program, and Professor of Pathology, UC Davis School of Medicine. His expertise is in veterinary pathology and infectious diseases.

Coenraad F.M. Hendriksen, DVM, PhD, is Head, Netherlands Centre for Alternatives to Animal Use, Chair, Alternatives to Animal Use at the Veterinary Faculty of Utrecht University, and Research Scientist, Netherlands Vaccine Institute. His expertise is in animal welfare concerns and because of his familiarity with lab animal issues in Europe.

William Morton, VMD, is Director, National Primate Research Center, University of Washington and Director of AIDS Research at the Regional Primate Research Center. His research interest is in retrovirology and he has published extensively on SIV variants and vaccine development. He is a well-known primatologist and has been an officer of the Association of Primate Veterinarians.

Emilie F. Rissman, PhD, is Professor of Biochemistry and Molecular Genetics, University of Virginia Medical School. Her expertise is in neurobiology and the effects of gender on behavior.

Randall J. Nelson, PhD, is Professor of Anatomy and Neurobiology and Executive Director of Animal Welfare and Compliance at The University of Tennessee Health Science Center. His research interests are in the dynamic control of hand movement and the effects of centrally-generated modulatory influences on somatosensory processing in the neocortex.

William S. Stokes, DVM, Dipl. ACLAM, is Director, National Toxicology Program, Interagency Center for the Evaluation of Alternative Toxicological Methods, Environmental Toxicology Program, NIEHS. He is also a captain in the US Public Health Service (USPHS) and Chief Veterinary Officer of the USPHS. His research interests are toxicological methods, including development, validation, and acceptance of new animal models and improved toxicological test systems.

Appendix D:
International Guiding Principles for Biomedical Research Involving Animals (1985)

INTRODUCTION

The International Guiding Principles for Biomedical Research Involving Animals were developed by the Council for International Organizations of Medical Sciences (CIOMS) as a result of extensive international and interdisciplinary consultations spanning the three-year period 1982-1984.

Animal experimentation is fundamental to the biomedical sciences, not only for the advancement of man's understanding of the nature of life and the mechanisms of specific vital processes, but also for the improvement of methods of prevention, diagnosis, and treatment of disease both in man and in animals. The use of animals is also indispensable for testing the potency and safety of biological substances used in human and veterinary medicine, and for determining the toxicity of the rapidly growing number of synthetic substances that never existed before in nature and which may represent a hazard to health. This extensive exploitation by man of animals implies philosophical and moral problems that are not peculiar to their use for scientific purposes, and there are no objective ethical criteria by which to judge claims and counterclaims in such matters. However, there is a consensus that deliberate cruelty is repugnant.

Suggestions had been received from several quarters that CIOMS, as an international nongovernmental organization representative of the biomedical community, would be ideally placed to propose a broadly based statement, acceptable worldwide in different cultural and legal back-grounds, and designed to create a greater understanding on the subject of biomedical research involving animals. Moreover, in several countries political action was being taken to stop or severely limit animal experi-mentation, and the Council of Europe had for some time been engaged in the elaboration of a convention to regulate the use of vertebrate animals for experiments or toxicity tests.

While many countries have general laws or regulations imposing pen-alties for ill-treatment of animals, relatively few make specific provision for their use for scientific purposes. In the few that have done so, the measures adopted vary widely, the extremes being: on the one hand, legally enforceable detailed regulations with licensing of experimenters and their premises together with an official inspectorate; on the other, entirely voluntary self-regulation by the biomedical community, with lay participation. Many variations are possible between these extremes, one intermediate situation being a legal requirement that experiments or other procedures involving the use of animals should be subject to the approval of ethical committees of specified composition.

In elaborating and publishing the International Guiding Principles the objective of CIOMS is not to duplicate such national regulations or voluntary codes as already exist but to provide a conceptual and ethical framework, acceptable both to the international biomedical community and to moderate animal welfare groups, for whatever regulatory measure each country or scientific body chooses to adopt in respect of the used animals for scientific purposes. The Principles strongly emphasize that there should not be such restrictions as would unduly hamper the ad-vance of biomedical science or the performance of necessary biological tests, but that, at the same time, biomedical scientists should not lose sight of their moral obligation to have a humane regard for their animal sub-jects, to prevent as far as possible pain and discomfort, and to be con-stantly alert to any possibility of achieving the same result without resort to living animals.

The International Guiding Principles are the product of the collabora-tion of a large and representative sample of the international biomedical community, including experts of the World Health Organization, and of consultations with responsible animal welfare groups. They have consti-tuted the agenda for three international meetings, the first of these being a Working Group that met in March 1983 to consider a preliminary draft prepared by CIOMS with consultant aid and the collaboration of the WHO Secretariat. The next meeting was the XVIIth CIOMS Round Table Con-ference, held in December 1983, to give the draft International Guiding Principles, as amended by the Working Group, a much wider exposure to criticism and suggestions. The third and last meeting, which took place in

June 1984, was of a CIOMS Expert Committee which met for a final review of the International Guiding Principles as revised in the light of comments made during the Round Table Conference and subsequently by correspondence.

The International Guiding Principles have already gained a considerable measure of acceptance internationally. European Medical Research Councils (EMRC), an international association that includes all the West European medical research councils, fully endorsed the Guiding Principles in 1984. Proposed U.S. Government Principles for the Utilization and Care of Vertebrate Animals used in Testing, Research and Training formulated in 1984 by the U.S. Interagency Research Animal Committee, were to a considerable extent based on the CIOMS Guiding Principles. In the same year, the Guiding Principles were endorsed by the WHO Advisory Committee on Medical Research at its 26th Session.

It is the hope of CIOMS that these Guiding Principles will provide useful criteria to which academic, governmental and industrial bodies may refer in framing their own codes of practice or legislation regarding the use of laboratory animals for scientific purposes.

Zbigniew Bankowski, M.D.
Executive Secretary, CIOMS

INTERNATIONAL GUIDING PRINCIPLES FOR
BIOMEDICAL RESEARCH INVOLVING ANIMALS

PREAMBLE

Experimentation with animals has made possible major contributions to biological knowledge and to the welfare of man and animals, particularly in the treatment and prevention of diseases. Many important advances in medical science have had their origins in basic biological research not primarily directed to practical ends as well as from applied research designed to investigate specific medical problems. There is still an urgent need for basic and applied research that will lead to the discovery of methods for the prevention and treatment of diseases for which adequate control methods are not yet available—notably the noncommunicable diseases and the endemic communicable diseases of warm climates.

Past progress has depended, and further progress in the foreseeable future will depend, largely on animal experimentation which, in the broad field of human medicine, is the prelude to experimental trials on human beings of, for example, new therapeutic, prophylactic, or diagnostic substances, devices, or procedures.

There are two international ethical codes intended principally for the guidance of countries or institutions that have not yet formulated their own ethical requirements for human experimentation: the Tokyo revision of *the Declaration of Helsinki* of the World Medical Association (1975); and *the Proposed International Guidelines for Biomedical Research Involving Human Subjects* of the Council for International Organizations of Medical Sciences and the World Health Organization (1982). These codes recognize that while experiments involving human subjects are a *sine qua non of* medical progress, they must be subject to strict ethical requirements. In order to ensure that such ethical requirements are observed, national and institutional ethical codes have also been elaborated with a view to the protection of human subjects involved in biomedical (including behavioural) research.

A major requirement both of national and international ethical codes for human experimentation, and of national legislation in many cases, is that new substances or devices should not be used for the first time on human beings unless previous tests on animals have provided a reasonable presumption of their safety.

The use of animals for predicting the probable effects of procedures on human beings entails responsibility for their welfare. In both human and veterinary medicine animals are used for behavioural, physiological, pathological, toxicological, and therapeutic research and for experimental surgery or surgical training and for testing drugs and biological preparations. The same responsibility toward the experimental animals prevails in all of these cases.

Because of differing legal systems and cultural backgrounds there are varying approaches to the use of animals for research, testing, or training in different countries. Nonetheless, their use should be always in accord with humane practices. The varying approaches in different countries to the use of animals for biomedical purposes, and the lack of relevant legislation or of formal self-regulatory mechanisms in some, point to the need for international guiding principles elaborated as a result of international and interdisciplinary consultations.

The guiding principles proposed here provide a framework for more specific national or institutional provisions. They apply not only to biomedical research but also to all uses of vertebrate animals for other biomedical purposes, including the production and testing of therapeutic, prophylactic, and diagnostic substances, the diagnosis of infections and intoxications in man and animals, and to any other procedures involving the use of intact live vertebrates.

1. BASIC PRINCIPLES

I. The advancement of biological knowledge and the development of improved means for the protection of the health and well-being both of man and of animals require recourse to experimentation on intact live animals of a wide variety of species.

II. Methods such as mathematical models, computer simulation and *in vitro* biological systems should be used wherever appropriate.

III. Animal experiments should be undertaken only after due consideration of their relevance for human or animal health and the advancement of biological knowledge.

IV. The animals selected for an experiment should be of an appropriate species and quality, and the minimum number required to obtain scientifically valid results.

V. Investigators and other personnel should never fail to treat animals as sentient, and should regard their proper care and use and the avoidance or minimization of discomfort, distress, or pain as ethical imperatives.

VI. Investigators should assume that procedures that would cause pain in human beings cause pain in other vertebrate species, although more needs to be known about the perception of pain in animals.

VII. Procedures with animals that may cause more than momentary or minimal pain or distress should be performed with appropriate sedation, analgesia, or anesthesia in accordance with accepted veterinary practice. Surgical or other painful procedures should not be performed on unanesthetized animals paralysed by chemical agents.

VIII. Where waivers are required in relation to the provisions of article VII, the decisions should not rest solely with the investigators directly concerned but should be made, with due regard to the provisions of articles IV, V, and VI, by a suitably constituted review body. Such waivers should not be made solely for the purposes of teaching or demonstration.

IX. At the end of, or, when appropriate, during an experiment, animals that would otherwise suffer severe or chronic pain, distress, discomfort, or disablement that cannot be relieved should be painlessly killed.

X. The best possible living conditions should be maintained for animals kept for biomedical purposes. Normally the care of animals should be under the supervision of veterinarians having experience in laboratory animal science. In any case, veterinary care should be available as required.

XI. It is the responsibility of the director of an institute or department using animals to ensure that investigators and personnel have appropriate qualifications or experience for conducting procedures on animals. Adequate opportunities shall be provided for in-service training, including the proper and humane concern for the animals under their care.

2. SPECIAL PROVISIONS

Where they are quantifiable, norms for the following provisions should be established by a national authority, national advisory council, or other competent body.

2.1 Acquisition

Specialized breeding establishments are the best source of the most commonly used experimental animals. Nonspecifically bred animals may be used only if they meet the research requirements, particularly for health and quality, and their acquisition is not in contradiction with national legislation and conservation policies.

2.2 Transportation

Where there are no regulations or statutory requirements governing the transport of animals, it is the duty of the director of an institute or department using animals to emphasize to the supplier and the carrier that the animals should be transported under humane and hygienic conditions.

2.3 Housing

Animal housing should be such as to ensure that the general health of the animals is safeguarded and that undue stress is avoided. Special attention should be given to the space allocation for each animal, according to species, and adequate standards of hygiene should be maintained as well as protection against predators, vermin, and other pests. Facilities for quarantine and isolation should be provided. Entry should normally be restricted to authorized persons.

2.4 Environmental Conditions

Environmental needs such as temperature, humidity, ventilation, lighting, and social interaction should be consistent with the needs of the species concerned. Noise and odour levels should be minimal. Proper facilities should be provided for the disposal of animals and animal waste.

2.5 Nutrition

Animals should receive a supply of foodstuffs appropriate to their requirements and of a quality and quantity adequate to preserve their health, and they should have free access to potable water, unless the object of the experiment is to study the effects of variations of these nutritional requirements.

2.6 Veterinary Care

Veterinary care, including a programme of health surveillance and disease prevention, should be available to breeding establishments and to institutions or departments using animals for biomedical purposes. Sick or injured animals should, according to circumstances, either receive appropriate veterinary care or be painlessly killed.

2.7 Records

Records should be kept of all experiments with animals and should be available for inspection. Information should be included regarding the various procedures which were carried out and the results of post mortem examinations if conducted.

3. MONITORING OF THE CARE AND USE OF ANIMALS FOR EXPERIMENTATION

3.1 Wherever animals are used for biomedical purposes, their care and use should be subject to the general principles and criteria set out above as well as to existing national policies. The observance of such principles and criteria should be encouraged by procedures for independent monitoring.

3.2 Principles and criteria and monitoring procedures should have as their objectives the avoidance of excessive or inappropriate use of experimental animals and encourage appropriate care and use before, during, or after experimentation. They may be established by: specific legislation laying down standards and providing for enforcement by an official inspectorate; by more general legislation requiring biomedical research institutions to provide for peer review in accordance with defined principles and criteria, sometimes with informed lay participation; or by voluntary self-regulation by the biomedical community. There are many possible variants of monitoring systems, according to the stress laid upon legislation on the one hand, and voluntary self-regulation on the other.

4. METHODS NOT INVOLVING ANIMALS: "ALTERNATIVES"

4.1 There remain many areas in biomedical research which, at least for the foreseeable future, will require animal experimentation. An intact live animal is more than the sum of the responses of isolated cells, tissues or organs; there are complex interactions in the whole animal that cannot be reproduced by biological or nonbiological "alternative" methods. The term "alternative" has come to be used by some to refer to a replacement of the use of living animals by other procedures, as well as methods which lead to a reduction in the numbers of animals required or to the refinement of experimental procedures.

4.2 The experimental procedures that are considered to be "alternatives" include nonbiological and biological methods. The nonbiological methods include mathematical modelling of structure-activity relationships based on the physico-chemical properties of drugs and other chemicals, and computer modelling of other biological processes. The biological methods include the use of micro-organisms, *in vitro* preparations (subcellular fractions, short-term cellular systems, whole organ perfusion, and cell and organ culture) and under some circumstances, invertebrates and vertebrate embryos. In addition to experimental procedures, retrospective and prospective epidemiological investigations on human and animal populations represent other approaches of major importance.

4.3 The adoption of "alternative" approaches is viewed as being complementary to the use of intact animals and their development and use should be actively encouraged for both scientific and humane reasons.